RETHINKING SCIENCE, TECHNOLOGY, AND SOCIAL CHANGE

RETHINKING SCIENCE, TECHNOLOGY, AND SOCIAL CHANGE

Ralph Schroeder

Stanford University Press
Stanford, California
2007

Stanford University Press
Stanford, California

Printed in the United States of America on acid-free, archival-quality paper

Library of Congress Cataloging-in-Publication Data

Schroeder, Ralph.
 Rethinking science, technology, and social change / Ralph Schroeder.
  p. cm.
 Includes bibliographical references and index.
 ISBN 978-0-8047-5588-7 (cloth : alk. paper)
 1. Science—Social aspects. 2. Technology—Social aspects. 3. Social history. I. Title.

Q175.5.S2996 2007
303.48'3—dc22

2007015603

Typeset by Thompson Type in 10/14 Minion

Contents

Preface

THIS BOOK HAS BEEN A LONG TIME IN THE MAKING, AND A FEW WORDS about how it came about may be useful. I first started thinking about the relation between technology and social change when teaching courses in this area at Brunel University and at Royal Holloway–University of London. At the time, I also began research on virtual reality technology and tried to draw on the relevant literature on the social implications of new technologies. Yet both in teaching and research, I found myself at odds with the prevalent notion that science and technology were socially shaped and seemed to do none of the shaping, at least in the sociological literature. This went against what I had learned from comparative historical sociology in my previous work about Max Weber and also in debates about rationality that I had encountered in philosophy and anthropology. So I began to develop an alternative position, drawing on Ian Hacking's early work, Ernest Gellner, Randall Collins, Richard Whitley, Stephan Fuchs, and others.

The opportunity to develop these arguments into a full-length book presented itself when I took a job at Chalmers University of Technology in Sweden. Teaching skeptical and instrumentally minded engineers about the social implications of science and technology, I soon learned, is quite different from teaching sociology students! I hope that the benefits of six years among the other of "The Two Cultures" at Chalmers is reflected in what follows. At the same time, I hope that the book will reflect my attempts to bring the sociology of the science and technology into mainstream sociological thinking, instead of remaining the subdisciplinary specialism it has become, staying almost completely isolated from mainstream sociological concerns.

The main arguments have been presented on a number of occasions to sociology of science and technology audiences with a mixed response: most have strongly disagreed with my challenge to the prevailing social shaping or constructivist orthodoxy. It has been interesting, however, that whenever I have presented these ideas, people have often said to me—usually backstage, after the talk—how much they agree with me. This book is one way to get the debate out into the open.

Although some of the ideas presented here have been presented in different form in journal publications, this systematic exposition will, I hope, provoke a lively discussion. In this vein, I have always thought that Max Weber's words are appropriate for academic progress: "Please polemicize as sharply as possible against my views on those points where we differ" (letter to Friedrich von Gottl-Ottlilienfeld, 14.4.1906). I am sanguine or realistic enough to know that social shaping and constructivism will be with us for some time. But over the long term, theoretical approaches in the social sciences come and go. And if the argument I make about how the advance of scientific knowledge works is correct, and if social science contributes to that advance, especially by opening itself up to what we know about broader historical patterns, then I believe that one day we will also have a thoroughly sociological understanding of science that recognizes how science is both part of—but also shapes—society.

Acknowledgments

I T IS A PLEASURE TO ACKNOWLEDGE MANY DEBTS: AT BRUNEL AND ROYAL Holloway, I learned much from discussions with Eric Hirsch and Ray Lee. When I started at Chalmers, I was fortunate that David Hounshell was a visitor during my first year and provided me with a solid grounding in the history of American R&D. Jan Hult, Marie Arehag, Bengt Berglund, Henrick Björck, and Jan Jönmark helped me over the years to get a deeper sense of the history of Swedish science and technology, not least by means of a number of memorable field trips to important historical sites. My PhD students at Chalmers, Ann-Sofie Axelsson, Ilona Heldal, and Maria Spante indulged the eclectic reading lists on my courses, and we had many hours of useful and enjoyable discussion. Colleagues at Chalmers also introduced me to the economics of innovation, and I am grateful for many discussions with Staffan Jacobsson, Maureen McKelvey, Linus Dahlander, and Annika Rickne. A visiting term at Stanford University's SCANCOR gave me the possibility to produce the first draft of this book and introduced me to new perspectives. My new colleagues at the Oxford Internet Institute have provided me with the opportunity to try out my ideas in the realm of new media. I have discussed the ideas over many years with Thomas Heimer, Leslie Haddon, Nina Degele, Richard Swedberg, and Alladi Venkatesh. The usual disclaimers apply.

My greatest thanks, as always, go to my family. They spend almost as much time as I do in the technologically mediated setting of the computer screen, but only my job, as my daughter memorably put it, consists of "blah blah blah"! I hope the reader will be more generous with me.

RETHINKING SCIENCE, TECHNOLOGY, AND SOCIAL CHANGE

1 Why Rethinking?

The Debates and the Argument

Two ideas about science, technology, and social change have dominated the social sciences for some time; both, I will argue in this book, mistaken. One is what I will call "speculative" scientific and technological determinism, the idea that science and technology cause wholesale changes in society. This is the view, for example, of those who argue that we live in an information or knowledge society, or that science and technology have revolutionized the society we live in today. It is also a view that is out of fashion among researchers, though it is widely held in society at large and often propounded in popular writings on science and technology.

The other is the idea that science and technology never determine social change in and of themselves or independently, but that this change is always already cultural or social. On this view, science and technology are inescapably part of a social context, and therefore no autonomous role in social change can be attributed to them. This view, known in the subdiscipline of the sociology of science and technology as "social shaping" or "social constructivism," is currently so well entrenched among academic researchers that it can be labeled an orthodoxy. This view also has a widely held counterpart in society at large, as in the saying that "it is never science (or technology) itself that causes change, it is people."

We will return to both views shortly and engage with them throughout this book. But for the reader who is tempted to reply immediately that the truth must surely lie somewhere between these two extremes, or that the problem must have been posed incorrectly to lead to such contradictory views, all I can say at this stage is that the formulation is not at fault, and that the answer that will be given here does not lie somewhere halfway between the two extremes. Briefly, my argument will be that scientific and technological determinism is true, but not in the wholesale way that (speculative) determinists argue or believe. Instead, determinism needs always to be yoked to the evidence about specific social changes that science and technology—independently—bring about. Put in a nutshell, the view I shall argue for here is that science and technology *do* determine social change, but from a social science perspective, their role in society is never independent of *what* they do to change the natural and social worlds.

This book will put forward several new arguments about the relationship between science, technology, and social change. I list the main ones here, and will elaborate the first two in the introduction and the others in later chapters, assessing—and amassing—the evidence to support them along the way.

1. Science, following Ian Hacking, is defined as "representing and intervening," and I add to this my definition of technology as "refining and manipulating." These are realist definitions that postulate that knowledge and the world are separate, as are artifacts and the environment they shape, which makes the main task of social science to analyze how the two sides interlock.

2. The social side of science, following Max Weber, is disenchantment, and technology extends this disenchantment into the social world by creating a cage of relations mediated by artifacts. Science and technology are cumulative, and disenchantment resulting from the growth of scientific knowledge is therefore progressive and displaces other forms of culture, while technology imposes an ever more powerful human footprint on the environment.

3. Science and technology are separate from culture, as well as from the political and economic spheres. Without this separation, it is impossible to grapple with the distinctiveness of modern science: a world-historically unique trajectory of cumulative knowledge that follows in the wake of the distinctive science-technology entwining in the mid-

nineteenth century and which, in turn, led to a unique pattern of sustained economic growth. The evidence for this analytical separation and these new patterns of social change must come from comparative history: macrohistorical comparison with premodern societies, pinpointing where and when the takeoff of science/technology and economic growth took place, and the more specific stages in which it subsequently did so.

4. Comparative history also provides two concepts that explain the distinctiveness of science and technology in the twentieth century—big science and large technological systems—which increase their scale and scope, and thus also increase their entwining with other institutions and their imprint on the environment. Recent history also allows us to chart the shift toward—and limits to—a global innovation system.

5. The most obvious impact of science and technology is via economic growth, but to gauge the significance of science and technology for everyday life, it is not enough to look at purely quantitative economic effects. Their advance must also be translated into the use of technologies in everyday life. Mass production and mass consumption have vastly extended the reach and volume of goods and services. The impact of technology—and more indirectly of science—on everyday life is thus to lead to a proliferation of technologically mediated activity and an ever more homogeneously diversified way of life.

In the introduction, I will elaborate several of these arguments about the relation between science, technology, and social change. No attempt will be made to provide a systematic review of other ideas about science, technology, and social change—these are readily available elsewhere—except to engage critically with them at various points in this book.[1] Still, it is worth saying a bit more about some key debates before plunging into the argument.

The theory of science, technology, and social change in the academic world has in the past two decades been dominated by developments at the forefront of social theory generally, in the 1970s and 1980s by social shaping with affinities to broader (neo-Marxist) debates about economic forces shaping society, and more recently by social constructivism, part of the larger trend of postmodernism in the social sciences. These currents will not be discussed in any detail except to highlight how the comparative-historical and institutional (or structural, if you prefer) alternative proposed here departs from them. But it is important at least to pin down the key ideas of these positions.

The argument of social shaping was directed against the "internalist" way of thinking about science and technology, which regarded scientific knowledge and technological innovation as a succession of ideas and improvements developing in isolation from and independent of their social contexts. This is similar to the way in which the history of culture and ideas as a series of interconnected and free-floating thoughts was challenged by more materialist conceptions (see Abercrombie, Hill, and Turner 1980). Against internalism, the social shaping perspective argued that science and technology could not be divorced from their social contexts, being shaped by dominant power interests. This perspective became known as the sociology of scientific knowledge (SSK) or science and technology studies (STS). Constructivism has continued this line of thought. The preoccupation with power has of late shifted toward a concern with meaning and identity in cultural constructivism, whereby science and technology are "suspended in webs of meaning that structure the possibilities of their action" (Hess 1997: 83) and are therefore not autonomous, objectively valid, or related directly to material objects. What is common to these positions, and what defines social constructivism, according to Hess (or cultural constructivism—constructivism can be used for both here) is that they are "studies that treat the social world as an exogenous, independent variable that shapes or causes some aspect of the content of science and technology" (Hess 1997: 82; see also Woolgar 1988). In short, for constructivism, as for SSK or STS, society shapes scientific knowledge.

These ideas were often first articulated for science, but, as already indicated in the quote by Hess, they came to be applied to technology as well. The idea of the social construction of technology was thus also developed from the earlier social shaping tradition.[2] If science and technology are always inescapably social, this goes against technological determinism, which is frequently criticized by writers in this tradition: "The technological, instead of being a sphere separate from society, is part of what makes society possible—in other words, it is constitutive of society" (MacKenzie and Wajcman 1999b: 23). We can note already that it is curious that technology is constitutive and yet not a separate sphere.[3] Note also that *technological* determinism is criticized—there is never, to my knowledge, talk of *scientific* determinism—though the latter seems to be part and parcel of the former. (Instead of *Rethinking Science, Technology, and Social Change* an alternative title for this book could be *In*

Defense of Scientific and Technological Determinism.) These arguments will be revisited throughout the book. For now, however, it will suffice to summarize the dominant perspective of social shaping and constructivism as arguing the opposite; namely, that science and technology are always already shaped by social and cultural forces or that they are inescapably social and cultural.

Apart from this recent postmodern or constructivist theorizing, the main contributions to the study of the role of science and technology have been made by historical and contemporary case studies of individual areas of science and technology (often informed by social shaping or constructivism). There have also been discussions in research policy about the social implications of new scientific discoveries and of individual new technologies. And finally, there have been high-powered debates about technology and quantitative measures of economic growth in economic history. We will draw on many of these later, but the problem with these more local, policy-oriented, and quantitative studies is that they only cover particular aspects of the science, technology, and society relationship. Put differently, what is lacking in these studies is an overall account of science, technology, and social change.

A related problem is that, apart from the sociology of science and technology, the social sciences have treated science and technology from within their narrowly disciplinary vantage points. Disciplines like philosophy—or the subdiscipline of the philosophy of science—treat the question of scientific truth as an epistemological issue, whereas others, such as anthropology, deal with science under the rubric of "rationality" in society. Or, to take another example, economics conceives of technology primarily as innovation and as a question of productivity or growth, while history or cultural studies may be more interested in the symbolic value of a technological artifact. To anticipate, I will argue that it is impossible to treat these questions separately; or, to put it the other way around, that it is necessary to answer the philosophical question, what is scientific knowledge?—and the sociological question of how science affects social relations—*simultaneously*. Similarly with technology, where innovation is one part of the impact and the everyday cultural significance another, and neither can be discussed without the other.

Further, and to anticipate a key argument that will be made at length, it is necessary to provide a comprehensive theory of the science-technology and social change relationship, *and* at the same time to take into account something that lies outside of this theory, namely, the (comparative-historical and other empirical) *evidence*. In other words, regardless of disciplines, the evidence for the role of science and technology in social change must be at a minimum comparative and historical *and* grapple with questions of macro-social change theoretically and empirically—rather than dealing with particular areas or microcontexts or resting on a priori philosophical assumptions.

The aim of the remainder of this chapter is to present definitions of science and technology and suggest how these can overcome some of the impasses in current debates.

Some Definitions

Before we can proceed with a definition, we need to ask two broad macrolevel questions: One is whether science and technology play a *unique* role in modern or industrial society, and the other is whether they have had an *autonomous* impact on society (as we shall see in a moment, the two are linked). The sociology of science and technology typically deals only with the second. The first has been tackled within a number of disciplines, foremost among them history (Smith and Marx 1994), economic history (Inkster 1991a; Mokyr 1990), and economics (Rosenberg 1982). The combination of the two questions also raises important issues in philosophy (Trigg 1993: 149–71) and anthropology (Horton 1970). Curiously, even though the same questions arise in different disciplines, there is very little interchange between them, and the sociology of science and technology has been quite insular in focusing on the second and not engaging with the first.

In economic history and comparative historical sociology, there is an emerging consensus that the role of science and technology in modern (or capitalist, or industrial) society has had unique social consequences or concomitants (this topic will be treated in more depth in Chapter 4). Regardless of whether the emphasis is on how science and technology foster industrial development (Inkster 1991a), or how they produce economic growth more narrowly conceived (Mokyr 1990), economic historians recognize that the effects of the Industrial Revolution in the nineteenth century were a watershed in the role of science and technology in society. Only at a particular point did science and technology *systematically* become oriented to the growth of knowledge and economic growth. And only in modern societies, to anticipate a point made by Collins (again, it will be developed later), does science become "high consensus rapid-discovery science" (1994: 157). The importance of this is that we can say that science and technology are not everywhere and at all times determined by the societies around them, but that instead, because of the consequences of scientific and technological advance for economic growth during this period, their relation with society changed too.

This much is not controversial—or at least, a variety of types of historical analysis would converge on this point if they were directly confronted with each other. There continue to be debates about the timing and the specific part played by science and technology in the Industrial Revolution (or the two industrial revolutions), to what extent the scientific revolution was a precondition of the Industrial Revolution, and so forth. These debates continue in several disciplines, the most advanced being in economic history (again, we will return to these debates, and to the unique role of modern science and technology, in Chapter 4). My point here is simply that from the point of view of comparative history, a distinctive trajectory is undeniable.

It is not possible to go directly from this comparative-historical argument to the autonomy of scientific and technological change. The further step is to say that sustained economic growth is a central feature of modern society that sets modern—or again industrial or capitalist—society apart from traditional or preindustrial or precapitalist societies (nothing in the argument here hangs on the three different labels, so I will use all three as appropriate in the context). In other words, both are unique, modern science and technology *and* modern sustained economic growth. If we now combine these two—the uniqueness of this type of economic growth and the unique growth of scientific knowledge and of technological development in industrial society—to say that there is a causal relationship between them, then there will nevertheless be an element of circularity in this argument.[4] For the purpose of the argument, what is required is only a "necessary condition" stipulation—one not without the other—since we may or may not be able to arrive at necessary and sufficient causality across the great divide between premodern and modern societies.

There may, however, be good reasons for this circularity: if it were not for the fact that scientific knowledge (and with it, technology of a certain type) could be separated from nonscientific belief systems in this way, it is difficult to see how any distinction between science and other kinds of belief systems could be made in the first place. Similarly, if the material basis of societies that have undergone the transformation of industrialization could not be separated from those that had not, there would be little point in setting modern or industrial societies apart at all.[5] Be that as it may, the implication of this argument is that a separation can be made between the social world and the world of scientific knowledge and physical artifacts, a separation which, as we shall see shortly, bears importantly on how the relation between science and technology and social change is conceptualized not only on this macro- and comparative-historical level, but on all levels of social scientific analysis—including, as we shall see, on the microlevel of everyday life.

The autonomy of science and technology follows at this point since their consequences, at least on this occasion, are different from those elsewhere. Yet now we need to ask, what is the significance of this autonomy? Marx thought that the main importance of science and technology lay in their capacity to transform the mode of production, but as MacKenzie has pointed out, this is too narrow since it leaves out, among other things, domestic technology or military technology (1984: 499, note 84).[6] In any case, if the autonomy of science and technology has been established by reference to its association with economic growth, then it should be the case that the consequences of this autonomy are not merely economic ones.

We shall soon come to the wider implications of these arguments; so far, all I have done is to argue that there is a consensus about the evidence for the uniqueness or distinctiveness of modern science and technology, and our definition must take this into account. But again, it is worth stressing that the "must" here has to come from the comparative-historical evidence *and* from the side of the conceptual or theoretical apparatus that we bring to the issue. So we must also ask, how should we define science and technology? What do they *do*?

Here it becomes useful to draw on Ian Hacking's discussion of science. Hacking contends that modern science "has been the adventure of the interlocking of representing and intervening" (1983: 146). "We shall count as real," he writes, "what we can use to intervene in the world to affect something else, or what the world can use to affect us" (1983: 146).[7] This idea can be extended to technology, except that in this case, we are dealing with physical artifacts rather than with knowledge since, as Agassi has pointed out, "at the very least . . . the implementation of any technique whatsoever involves both physical and social activities" (1985: 25; cf. Bimber 1994: 88). Or, as Price puts it, "if one wishes to do something to something, what one uses is a technique rather than an idea" (1986: 240). In other words, technological artifacts are where the human and the natural or physical environments meet, but technology always involves (physical) hardware. Paraphrasing Hacking's conception of science, we can say then that modern technology has been the adventure of the inter-locking of refining and manipulating since technological advance consists of the process whereby artifacts are continually being modified in order to en-hance or extend our mastery of the world.[8] Science is directed at the natural or physical world, technology at the physical environment of human beings.

This is what science and technology, respectively, *do* to the natural and physical world and to the natural and human environment. This adds pragma-tism to Hacking's realism. But *what* they do simultaneously has social effects. Max Weber's ideas can take us further here. What science and technology do on the social side is a disenchantment by more powerful knowledge and a "caging" into our uses of more effective tools: more powerful knowledge adds to and displaces other beliefs, while enhanced tools add to and complement our existing range of tools. Weber regarded science and technology as central to the process of disenchantment, or the increasing extension of instrumen-tal rationality throughout the social world (Brubaker 1984: 29–35; Schroeder 1995), which simultaneously creates an "iron cage" of instrumentally rational institutions.[9] Weber was a cultural pessimist about this process; Gellner offers a corrective when he calls this a "rubber cage," which is much more user-friendly and reenchanted with consumerism than Weber anticipated (1987: 152–65). Moreover, caging is a somewhat misleading and limited metaphor—an "exoskeleton," a cage that *serves* human beings, may be more appropriate since the advance of science and technology also gives us greater power over the environment, extending the human footprint.

Weber's notion of disenchantment pertains to modern or industrial so-ciety generally, and it specifies a pattern that accompanies all scientific and technological change within this type of society. Thus, there are always two sides to the advance of science and technology: on the one hand is the advance of instrumental rationality, or of seeking the most efficient means to achieve a given end, which entails an increasing mastery over the natural and social worlds; on the other, this process also brings about the increasing impersonal-ity of the external conditions of life. The consequences of scientific and tech-nological advance are therefore not just economic ones; they apply to all areas of social life.

This conception of science and technology enables us to identify the con-tribution that specific advances in scientific knowledge and technological artifacts make to the process of disenchantment since it allows us to say what gains have been made in each case by instrumental rationality. The notion of scientific and technological *advance* (used here in a purely neutral sense) therefore entails a "realist" position inasmuch as it rests on the notion that scientific knowledge is separable from the world and that artifacts are physical objects, and it simultaneously takes into account the effect of this advance on the social world by means of Weber's concept of disenchantment.

The central concern of the sociology of science and technology thus becomes how we can translate the one into the other; that is, to translate the ways in which knowledge intervenes in the world and artifacts manipulate it—into the ways in which the external conditions of social life become increasingly governed by how knowledge and technology are deployed. Wherever we encounter the science, technology, and society relationship, we must be able to identify both an advance in representing/intervening and/or refining/manipulating that has taken place—and how social relationships have changed in accordance with more powerful knowledge and more effective artifacts.

This realist approach to science and technology applies to all levels of social change—macro-, meso- and micro-. Once a distinction has been made between science and the rest of the social world, or between a sphere of cognition whose validity is independent of social life (realism tells us only that knowledge is separate from the world and that artifacts are physical, not what the significance of this separation is), then we can operationalize realism by identifying the separate impact of this realm of knowledge and of physical artifacts on society.

This is a good point at which to spell out the difference between the realist and pragmatist argument made here and the social shaping and constructivist approaches. What social shaping and social constructivism leave out, and what sets the position put forward here apart, is the coupling between science and technology and the physical or natural world. This coupling transforms society independently of social forces. This is illustrated in Figure 1.1, where social shaping and constructivism recognize only the relationships above the dotted line, whereas the position put forward focuses on the relationship between science and technology and the physical/natural world below the dotted line (indicated by the arrow on the left), and thus recognizes the independent impact of scientific and technological advance on social change (indicated by the arrow on the right).

Note, first, that this latter position does not exclude the arrow above the dotted line, but regards this relationship as secondary. Note, secondly, that I argue for this determinism only insofar as it can be yoked to specific social

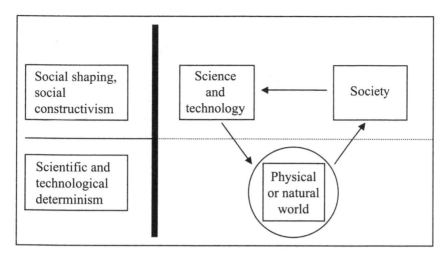

FIGURE 1.1 Social shaping and social constructivism contrasted with scientific and technological determinism.

processes (to avoid speculative determinism); in other words, insofar as the two arrows that cross the dotted line can be given a concrete content and sociological significance. Note finally that even if the arrow going from science and technology to society above the dotted line is seen as two-way (such a two-way process is often implied, but not spelled out, in social shaping and constructivist theories), this still does not take into account the relation to the physical and natural worlds, as here. The realist and pragmatist position put forward here implies an epistemological conception of knower and what is known as being separable, and of an analytical separation between the physical/natural worlds and the social worlds, even if there is an increasing interlocking between them in practice.

To this realist and pragmatist approach it will be necessary to add—again, it has already been alluded to and we will return to it later—that the relation between science and technology has been variable. Until the mid-nineteenth century, the two were not closely linked (Collins 1986: 113), but in the mid-nineteenth century, high-consensus rapid-discovery science became linked to technology. Cowan goes so far as to say that "in the twentieth century, it has proved very hard to distinguish between technology and science. For most of this century, technological development has been conducted using scientific methods and scientific research has been conducted, and funded, for technological reasons" (1997: 221). As mentioned earlier, this intertwining has not been systematically theorized for the science-technology and society relationship.[10] Here, the best I can do is to relate this to my definitions of science and technology, which means in this context that we should see both processes, of representing and intervening, refining and manipulating, and the disenchantment and caging attendant upon these, all working increasingly in tandem. Therefore, although science and technology predate modern society, and high-consensus rapid-discovery science led to the takeoff of economic growth in the course of the Industrial Revolution, it is only possible to speak of a progressive and systematic advance of science and technology and of disenchantment affecting society *more widely* from the time of this link onward.

The further intertwining of science and technology in the twentieth century will be described further in Chapters 2 and 4. In Chapter 3 we will encounter another feature of twentieth-century science and technology, which is that science has on a number of occasions become big science (Galison and Hevly 1992; Price 1986), as in the cases of particle physics or the human genome project, while technological artifacts have become part of "large technological systems" (Hughes 1987), as, for example, with electrification and telecommunications. In these cases, it is necessary to tackle simultaneously the wide-ranging social implications of science and technology on the one hand, and how they have focused the attention of a large part of the research and development community and required a large-scale mobilization of social resources on the other. This means, too, that in these cases, the examination of the disenchanting consequences of scientific and technological advance must encompass a wide range of simultaneous developments. And we will need to remember that these labels only pertain to certain types of scientific and technological advance. Others, like smaller-scale laboratory research or stand-alone domestic technologies, will require different points of departure.

The relationship between science and technology and social life is thus always a question of levels, and it is important to go beyond both a particularism that is unable to arrive at any general patterns of social change (as in studies of individual areas of science and technology), as well as beyond abstraction that does not apply to any concrete contexts: for different kinds of scientific and technological advances it is necessary to address different aspects of the interrelationship between the two sides, but it is also necessary to see the links between them. As we shall see, in the case of big science or large technological systems, for example, it may be necessary to focus on the institutional momentum that has built up behind the research and development efforts. For consumer electronics, along similar lines, it may be necessary to examine the feedback loops between research on new devices and the domestic contexts in which they are used. In any case, the later chapters in this book will identify several such patterns.

No doubt there are many different ways in which scientific and technological advances translate into everyday settings and it may be impossible to determine, a priori, whether science and technology or social forces do the shaping. What *is* clear from the outset, or what follows from the definitions given here, however, is that there will always be two sides to this interplay; the side of an ongoing adventure (or advance, in my terminology) of representing and intervening, or of refining and manipulating, and the side of disenchantment, or of an advance in instrumental efficacy and of the depersonalization of the external conditions of life—by means of greater control over and more mediation with the environment.

Beyond Social Shaping and Constructivism, and Some Puzzles Resolved

Before we go any further, some puzzles or seeming contradictions that follow from my definitions can be anticipated. I will argue that one of the keys to understanding the relationship between science, technology, and social change is to recognize that science is in crucial respects separate from society and from culture. A common response to this idea is: How is it possible to argue that science is separate from society? Isn't everything social, made by people? And don't all ideas or beliefs have to be part of culture?

This position only has to be put in a negative form—science and technology can never be anything but social or culturally shaped or constructed—to recognize that there must be something wrong, too, with the idea that science must be social or cultural through and through. Science is indeed social, but it *must* also be independent of society since it clearly imposes constraints *on* us—for example, when scientific laws are valid in relation to how they pertain to the physical world, and thus regardless of whether society or culture shapes or constructs them so. (We will identify some other constraints later.) In a similar way, science can indeed be regarded as part of our modern culture, but it must also be possible to separate science from culture since there is clearly a difference between science and other things that social scientists want to refer to as culture.

Since these are issues that go the heart of the argument of this book, it is worth spending some time on them here. One way to do this is by briefly focusing on the notion of an "essential" difference between science and culture. One of the most interesting recent books in the sociology of science and culture is titled *Against Essentialism: A Theory of Culture and Society* (Fuchs 2001). It argues that the distinction between science on the one hand—and culture and society on the other—is false. In this, the book shares much with recent constructivist ideas in the sociology of science and culture. But *Against Essentialism* makes an *objectivist* case for this inseparability: Fuchs argues that it is possible to provide an objective social scientific account of culture, *including* science, in terms of its network structure. *Against Essentialism* argues that science is not essentially different from other parts of culture, but only in so far as its network is harder than other parts of culture.

Let us retrace this argument in his own words: Fuchs (elsewhere) argues that science is cumulative: "what makes a science scientific . . . is its high instrumental and experimental capacity for progress" (2002: 33), which—in my argument here, but not for Fuchs—makes it unlike other areas of culture. For Fuchs, the hardness of this part of culture can be explained as follows: "Rationality prospers when the relevant world has been simplified and quantified, concentrating the attention space on a small and domesticated set of well-understood variables and parameters" (2001: 137). This seems like a very local explanation of rationality, but it only takes a moment's reflection to understand why such a concentrated effort yields high-consensus rapid-discovery science, or blazes a trail of knowledge that is more universal than other, say philosophical, efforts that attempt to tame issues that cannot be so simplified and quantified.

But let us follow Fuchs further: he goes on to argue, "what makes a culture 'hard' and realist, rather than 'soft' and constructivist, is hardware, among other things" (2001: 306). "Realism," he says, "increases when a culture is grounded in routine machines, tools, and instruments, around the formal and technical cores of organizations. This effect is strengthened further as the material means of culture are monopolized by an organizational hegemony . . . In laboratory sciences that occasion more copresence, encounters and groups, realism is anchored in the tangible reality of a here and now, with its physical interventions and manipulations" (2001: 330). Again, the terminology has similarities to the definitions of science and technology presented earlier, and it can easily be seen that this is different from other nonrealist parts of culture that we can think of. Fuchs thus describes the realm of science and technology as being quite different from other parts of culture. His argument is therefore misleading when he describes the realist part of networks in such local terms: copresence and anchoring in the tangible reality *here* and *now* may be required for the *local* production of scientific and technological advance, but what is unique about this part of culture is how easily it can be transferred to other places, and thus how context independent it is.

My argument against Fuchs is that this difference entails that science and the other parts of culture are not just two parts of the same animal, so to speak, but that they are different animals altogether—an essential difference. I would argue that analytically, but also from the point of the view of the evidence, there are only two options: one is to separate science from culture altogether, and the other is to subsume science under the rubric of culture, but to say that in this case science consists of an essentially different part of culture.[11] These two options are represented in Figure 1.2, where the triangular wedge of science/ technology can fall within the circle of culture but be analytically separable and occupy an increasing amount of space within it, or it can be separate from the circle of culture in the first place. (Sociological analysis has far less trouble, if any, with the separability of the other two spheres, of politics and economics.)

We will come back to this repeatedly, but the issue cannot, of course, be resolved in the end merely on a conceptual level. I will have to show later, in practice or by reference to substantive social changes, the extent to which science/technology *translate* into cultural (or into political and economic) change, having been separable from it—or vice versa. Be that as it may for now, in what follows I shall agree with Fuchs that this hardness of science and technology, as opposed to other parts of culture, needs explaining.

There is a related puzzle: Ideally, we should be able to treat scientific knowledge as a belief system. If science is "our" belief system—or culture or ideology—then one approach has been to argue, as Gellner does, that all beliefs should be translatable into another language (1979). So, for example, religion can be translated into economic benefits: salvation payoffs, premiums for deferred gratification, and the like. A related approach, proposed by Collins (and which will be pursued below), is to treat scientific truth or knowledge as a "sacred object of the scientific cult" (1993: 302), with rituals designed to reinforce this deity: common worship of truth, status deference in the order of the scientific priesthood, and so forth. The reason for mentioning these two arguments is to notice their limitations. As Gellner and Collins recognize, treating science as a belief system can only ever partly work for science, because science also locks onto the world and thus changes the world in ways

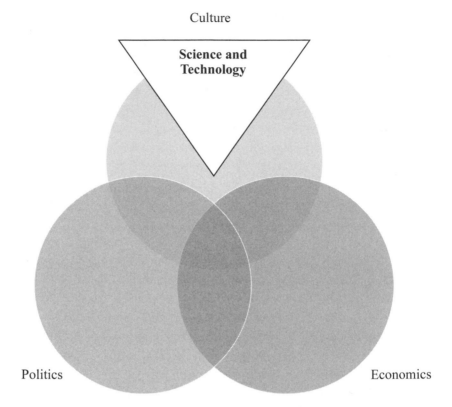

FIGURE 1.2 The relation between the spheres of culture, politics, economics, and science and technology.

that other belief systems do not (Collins 1975: 520; Gellner 1988: 70–90). Put differently, our belief system *can* be explained in this way, as ideology or as culture, or as the worldview that dominates our society. Yet the sociological significance of science does not lie primarily in the fact that it can be seen as a belief system like others, but in what scientific knowledge *does* as a unique belief system, and this, again, is to transform the world, part of which, as we shall see, is to eliminate other belief systems or all-encompassing worldviews.

Something similar applies to technology. The sociology of technology, and also other social science conceptions of technology, have been unable—bar some notable exceptions that we will return to—to make up their minds whether technology must consist of material artifacts: Should technology be restricted to hardware, to machines? No, it is typically argued, technology is always already cultural or social, and therefore necessarily also consists of ideas (as constructivists claim), or it includes organizational innovations apart from artifacts (this is often argued in the economics of innovation), and the like. Again, I will depart from this consensus in arguing that technology must at a minimum include a material component, and that the sociologically significant component of technology apart from its artifactual nature is the part of the environment that has been transformed by this technology.

The Extension of the Human Footprint

A different way to clarify the separability of science and technology from culture and society is to pursue the puzzle that science and technology are produced by human beings: how, it can be asked, is it possible for science and technology to be outside of a social context if science and technology are obviously human products and therefore inevitably social? But to paraphrase Marx, we may make science and technology, but we cannot do so as we please. Science also constrains us, as when the evidence compels the validity of knowledge, and technology constrains and enables our relation to the environment around us (and vice versa).

Another argument can be brought to bear, not from the side of concepts or theory, but from the side of the evidence. If scientific and technological determinism is true only in relation to its social context, the social context is literally universal in the following way: elaborating on the term borrowed from environmentalism earlier, which talks, in the context of sustainability, about the human footprint in the environment, we can use this concept in a broader sense, to indicate the transformation of the environment by science and technology (the connection with the definitions of science and technology in terms of disenchantment and the notions of caging and a human exoskeleton will be obvious here). If so, it can be recognized immediately that this human footprint has grown ever larger, reaching a historically unprecedented level of growth in the twentieth century that has encompassed the globe—and beyond. As McNeill puts it, "the human race, without intending anything of the sort, has undertaken a giant uncontrolled experiment on the earth . . . Although there are a few kinds of environmental change that are new in the twentieth century . . . for the most part the ecological peculiarity of the twentieth century is a matter of scale and intensity . . . matters that for millennia were local concerns became global" (2000: 4).

This footprint may be deepest in the most built-up areas of human habitation, but the environment we have transformed also lies partly outside of all human habitation, and thus of society: it has extended into realms large and small that lie outside of social relationships, only interlocked by knowledge or by machines (for example, by instruments of measurement or observation). Among the examples that come to mind are ocean depths and galaxies on the macrolevel and subatomic particles and genetic material on the microlevel.[12] To the objection that the human footprint must be social because it is created by humans, it can therefore be replied that not all the extensions of human activity are within the compass of social science, and this especially applies to (physical and extrahuman) phenomena covered—or interlocked—by science and technology and their instruments. Thus the impact of science and technology, insofar as the human footprint extends beyond society in any meaningful sense, is literally universal.

Again, the same seeming contradiction crops up in relation to technology: the argument made here is typically countered with: there is no such thing as technology without humans, or technology outside a social context! I will argue against this too, though it is necessary to make an analytical separation that is somewhat different from the case of science. What lies outside of science is the physical world or nature, the reality in which science represents and intervenes; what lies outside technology is the physical environment to which humans are subject, and which in this context is nothing apart from how this environment undergoes an ongoing refinement and manipulation by artifacts. (Refinement, incidentally, might be taken to imply making better. I use *refining*, like *advance*, in a value neutral way: a cage can be "refined" by making it more constraining, just as it is possible to "advance" to nuclear Armageddon.) In short, science and technology, though analytically separate, are never separate in practice from what they *do* to the (physical or natural) world, and this is also the key to resolving the seeming contradictions that have just been mentioned and advancing beyond the mistaken nothing-outside-of-the-social-context view.

Proponents of social shaping, and even more so constructivists, want to blur the distinction between science and nature, or between culture and nature, and partly for this reason they cannot identify the sense in which scientific knowledge grows and changes the natural or physical world. Similarly with technology which, if it is only culture, only changes our beliefs or ideas, rather than the external environment. A-social scientific realists, on the other hand, want to isolate truth from the world, and have a similar problem in not being able to specify any concrete changes that science makes in the world. Instead, they dwell on truth in the realm of abstract ideas—the internalists that social shaping rightly criticized. And similarly for a-social technological determinists who claim that technology changes everything, which entails that technology does not actually change any concrete social relations. (But again, in view of the premises of my argument that it is essential to show what knowledge and artifacts *do,* and not just assert their universality in the abstract—the autonomy from the social context, and diffusion throughout various social contexts, will need to be demonstrated *in practice* in what follows.)

To get beyond these abstractions, we need to take the evidence into consideration, since all our concepts are bound or bounded by evidence. The best evidence we have for how the relation between science, technology, and society has changed comes from comparative history and substantive sociological findings, and it is to these that we can now turn.

Before we do so, a brief map of the book is called for. Chapter 2 examines the institutional bases of scientific and technological advance; how are scientific disciplines and technological artifacts organized to move forward rapidly? This process has so far mainly been described for individual cases, but here an attempt is made to cover how these institutions are organized to foster advance along a cumulative frontier and how they draw on resources from society and establish their legitimacy. In Chapter 3, we will move on to the main manifestations of these institutions in the twentieth century, big science and large technological systems. These two large-scale institutions have become phenomena that reach out beyond science and technology and affect society at large. Our dependence on these two institutions has become taken for granted, but they only became possible in a society in which economic growth had for the first time become a permanent and routine feature. The coupling of technology and mass production changed the scale and scope of consumption, and how a steady stream of innovations are turned into mass consumer goods has been much discussed in economic history. In Chapter 4 this process is described from a longer-term historical perspective: what are the main stages in which this process became institutionalized on a large scale, and when and how did it become widespread throughout the developed world?

The study of how innovation is driven (or not) by demand in today's economy on the production side has been much debated. The *uses* of technologies, on the other hand, have been relatively neglected. Chapter 5, the first chapter to address this topic here, examines information and communication technologies as a large technological system and their impact on politics. The cumulative and systematic impact of technologies in relation to politics has often been overlooked in relation to political change, as has the fact that national systems have in an important sense converged. Similarly with the consumption of technologies changes everyday life, a topic that has not been extensively studied, and hardly at all from a long-term and comparative perspective. Chapter 6 therefore examines three such technologies in detail; car, telephone, and television. These, it is argued, have changed our everyday lives in the direction of a more homogeneously diversified lifestyle and culture. The conclusion argues, finally, that these substantive ideas about science, technology, and social change add up to a new theoretical agenda that goes beyond social shaping and constructivism and a crude or speculative determinism. It spells out the implications of this agenda, the extent and limitations of how it has been worked out in this book, and the consequences for wider debates about how science and technology change society.

2 The Social Organization of Scientific and Technological Advance

The Scope and Limits of Science and Technology in Society

Before delving into the social organization of science and technology, it is worth recalling (as argued in the first chapter) that science and technology are also extrasocial. This is because the relationships between how the world is represented and the physical world itself, and how artifacts intervene in the physical environment, obtain regardless of social context. The enhanced "interlocking" of representations and intervention in the physical world—and the introduction and addition of ever more powerful new artifacts for the manipulation of the environment in the case of technology—are extrasocial in the sense that they apply across a range of social contexts. This characteristic of scientific and technological advance is therefore partly related to the fact that the relationships between scientific knowledge and technological artifacts and the physical world obtain everywhere in the same way and partly to the fact that these interlocking relationships can be reproduced in any social setting. Here we will therefore need to trace how the consequences of the advance of science and technology in relation to the physical environment simultaneously gain their increasingly extrasocial (in the sense of nonlocal) position throughout the social world. In other words, social science is concerned with what science and technology *do*, and it is the increasingly universal (in the sense of being widespread throughout the social world) practical effect of this interlocking that we will be concerned with in this chapter.

The first place to pursue this extralocal advance is in the production of scientific knowledge itself. Cole has distinguished between "local knowledge outcomes" and "communal knowledge outcomes," and says: "I accept the fact that local knowledge outcomes may be influenced by social processes and chance factors, but I do not believe that anyone has demonstrated that the content of communal knowledge outcomes is influenced by social variables and processes" (1992: 29).[1] How these communal knowledge outcomes emerge is therefore something that must be explained. To do this, we can follow Collins and Fuchs, who build on Cole's (1992: 15–17) distinction between the "core" of scientific knowledge—the consensus that is accepted as given by scientists and used as a starting point—and the "research frontier," which may or may not become part of the consensus as scientific knowledge advances.

Among the groups or organizations at the research frontier, or for "science-in-the-making" as it is also known, we find competition and controversy, while for the core, or "science-already-made," we find cumulation and consensus. "The scientific community," writes Collins, "is oriented toward the moving edge of truth, a frontier that expands outwards while preserving what has already been brought inside. It is a cult of accumulated discovery" (1993: 303). It is therefore possible to agree with Fuchs's argument that the workings of science (and, I would add, technology) do not need any philosophical or epistemological foundations (1992, see also Fuchs 2002), and social science therefore does not need to enter into these stormy waters of philosophy. What needs to be shown is only how advancing knowledge gains its strength—or hardness—in the social world. In this respect, the argument chimes with recent constructivist accounts of science. Where I disagree with Fuchs (and with constructivism), as mentioned earlier, is that he also thinks that science has no "translocal foundations for knowledge" (1992: 12) or, again, that the foundations of science are continuous with the bases of culture. This takes constructivism too far, not for philosophical reasons but because the evidence for the translocal (social and physical) advance and diffusion of scientific knowledge (Collins's "accumulated discovery"), and for a legitimacy of science that is separate from the legitimacy of the rest of culture—as we shall see—is overwhelming.

Against "universality," it is, of course, possible to identify limitations of scientific knowledge, or places where the research frontier is stalled or embattled. So, for example, there are evidently certain areas where many would put limits on science as a "belief system," even though, as Gellner has argued, science *in general* is *the* exclusive legitimate mode of cognition of industrial society.[2] And, as Collins (1993) has pointed out, social science shares with natural science the *aim* of producing cumulative knowledge and so they are part and parcel of the same mode of cognition insofar as the former fits the definition of science presented in the first chapter. Another limitation—as we shall see—is that the scientific enterprise on the research frontier competes for resources with other organizations. A different way to describe these limitations is to put them into sociological language: if science is a functional requirement of industrial society, we should also be able to recognize "deviance" or "pathology," or perhaps "conflict" in addition to the "cohesion" of legitimate knowledge in society. We can no doubt find many examples and ways of specifying such contested boundaries of science, but the point of bringing up these limitations here is that deviance also allows us to recognize one way in which science will remain universal and unchallenged indefinitely; namely, in the sense that the absence of science as a social institution is literally (social scientifically and otherwise) unthinkable—unlike other social institutions such as the state or the market.

This point, again, applies in a somewhat different way to technology. To recognize this, we must briefly follow Collins: One way to describe the social impact of science and technology, according to Collins, is to conceptualize it as the outward flow of laboratory equipment into the world. What is new, from about 1600 onwards in Europe, he says, is "high-consensus rapid-discovery science," or "secure knowledge" and "a train of new results" (1994: 157; see also 1998: 533–38). "What was discovered," he continues, "was a method of discovery; confidence was soon built up that techniques could be modified and recombined endlessly, with new discoveries guaranteed continually along the way. And the research technologies gave a strong sense of the objectivity of the phenomena, since they were physically demonstrable. The practical activity of perfecting each technique consisted in modifying it until it would reliably repeat the phenomena at will" (1994: 163). Note that this fits well with the ideas borrowed and developed from Hacking and Weber and that Collins (and Hacking and Weber and I) scrupulously avoids an "idealist" account of science and technology: "advance" here takes place only in relation to the physical and social worlds conjointly and never just in the realm of ideas. Collins goes on to describe a process whereby there is an "outward flow of lab technology," and "laboratory technology" is "exported into the lay world" (1993: 315): "One machine gives rise to another in a genealogy of succession: by modifying the past machine, or by cloning it from another in the same laboratory, or by a kind of sexual reproduction recombining parts from several existing pieces of equipment" (1994: 164).

Now, along with Collins, we can follow these processes even further to their endpoint, to the *use* of innovations or artifacts in the lives of consumers (a trail which we will need to pick up later in this book): "After modifications at the hands of scientists, the equipment may become commercially viable when reintroduced into the lay world . . . Once this happens, the research process is legitimized to a high degree: not merely on the level of ideology (which may wax and wane), but in the taken-for-granted practices of everyday life. Rapid discovery science generates a strong sense of its objectivity and factuality because its research technology produces many allies" (1994: 165; see also Shinn and Joerges 2002). Put differently, "in return for support, laypersons receive material benefits, and often these take the form of equipment that can be individually owned and used, giving huge numbers of persons a visible stake in the scientific research enterprise" (1994: 170).[3]

This sequence from how new research technologies generate discovery to how they become spread out and taken-for-granted in everyday life can be tied to the overall argument being made here: in the recent sociology of science and technology, there are typically accounts about how ideas or beliefs go into the shaping of new knowledge or new technologies or how beliefs about new knowledge or new technologies are the most sociologically significant feature about them. Similarly, as we shall see later, in economics, where innovation is a response to "choice" or "demand." These are arguments about local contexts and culture—or, in economics, abstract invidual choosers or maximizers. Here, in contrast, the focus is on processes whereby knowledge that is embodied in artifacts (including research apparatus) proceeds from the lab and into the market and ultimately into the home. (Or—to use a somewhat different illustration that makes clear that the argument is not specific to home consumer technologies—the medicines that go from the lab into the hospital and ultimately into the patient's body.) My argument is that this is a more concrete way of explaining the role of science and technology in society than to look at the culture or beliefs among its creators and receivers.

To return to "limits" then, we can conduct a thought experiment about the universal nature of technology: to be sure, it is possible to conceive of this outward flow as waning or discontinuing, to see new technologies being removed from social life for their undesirable effects, and to envision the boundaries of the social impact of science and technology being pushed back and resulting in a flow back into more self-contained areas, perhaps into the laboratory where laboratory equipment is used only for its own sake. But again, much of this declining or disappearing impact of the role of science and technology in society could only be achieved by technological means.

To summarize the first part of the argument: on the most macrolevel, scientific and technological advance are universal in the sense that they define industrial society—not in the abstract, but in terms of how they have transformed that society through an enhanced interlocking with the physical environment. Once we move from this general level to the mesolevel of institutions *within* industrial society, we will need to see in what follows how this universality is embodied in scientific knowledge and technological artifacts that fill the available space throughout society—a lesser form of "universality." Before we can do so, we must first examine the changing shape of the organizations that produce scientific and technological advance.

The first point to notice on the mesolevel is that the organizations that promote scientific knowledge and technical expertise have been gaining at the expense of other organizations. Within the educational system and within the research community, certain types of knowledge claims—and their carriers, scientists and engineers—have advanced their monopoly on valid knowledge, and these claims have established themselves ever more firmly in relation to other forms of "knowledge" (the scare quotes indicate that this is not legitimate knowledge) such as religion or traditional medicine—and their carriers, such as priests or healers (Drori, Meyer, Ramirez, and Schofer, 2003: 197, 210 and passim).

If we pursue this mesolevel further, one question that has received a lot of attention is: How have different scientific disciplines established their autonomy from other branches of knowledge? How have they been able to define their content, police their boundaries, and secure their status as professions? What this focus on the competition *between* scientific disciplines overlooks is the contrast with other, nonscientific forms of belief or of knowledge, and this can only be done by setting scientific organizations apart from other organizations. As Whitley puts it, "modern sciences . . . attempt to monopolize the production of true knowledge about the world" (2000: 1), and he sees the sciences as work organizations that differ from other work organizations in that "they institutionalize the dominant value of producing new knowledge which goes beyond, and is an improvement on, previous work" (2000: 11). This definition meshes well with the definitions of science and technology given earlier and can provide the basis for the discussion of the specific organizational forms in which knowledge (including scientific and technological advance) is produced—which is the subject of this chapter. And unlike Fuchs, Whitley accepts that this definition sets "the sciences apart from many"—I would go further and say from *any*—"other areas of cultural production" (2000: 12).

The idea that scientific progress can be understood as the "extension of certified knowledge" (Fuchs 1992: 3), which is the functionalist, normative Mertonian approach in the sociology of science, has been much criticized. The main criticism, however, has been that this approach seems to ascribe to science some particular normative implications or elevate it into an unchallengeable truth. On the argument presented here, this philosophical debate is avoided altogether: the "extension of certified knowledge" has to be understood sociologically, in terms of how knowledge *works* in society, and in this sense the Mertonian view remains useful; that is, an extension of certified knowledge is taking place, without putting a positive evaluation on this extension, and without prejudging whether this working of knowledge may also on occasion go awry.

Another way to pinpoint the separateness of science as an institution is to take issue with Turner, who argues that science (and medicine) is not among the main or "core" institutions in modern society—unlike education, law, and economics—because it is not as autonomous as these other core institutions (1997: 8). Yet, as Weingart points out, science is possibly *the* fastest growing institution in modern societies (2003: 36). And if my argument is correct, science is *more* than a single autonomous institution; it is among the three spheres of society! Going from macro- (science as one sphere of society and an exclusive mode of cognition) to meso- (with the sciences as individual organizations, even if they share a common goal) will inevitably entail a shift, but one that the schema elaborated here allows us to accommodate. Thus we will come closer to an accurate picture of the sciences when we proceed now to identify in what sense science is more autonomous than these other institutions.

Now it is true that, as a sphere affecting our *everyday* lives, and among the mesolevel institutions, science is only one among many institutions. And in another sense, it is only a *part* of other institutions; education, law, the economy—in each of these institutions, science plays a part. Here then is another paradox: science at the phenomenological or everyday level and the mesolevel of institutions appears to be only a small part of society, but at the macrolevel, and in terms of its consequences, it dominates and pervades society—it is our privileged form of "belief," and again, transforms wide swathes of our human and natural environments beyond recognition. But saying "wide swathes" is not specific, and we can once more take issue with what I earlier called speculative technological (or scientific) determinism, claiming that science and technology change "everything." It is easy to insist on this overall, large-scale importance of science and technology; it is more difficult to specify its precise boundaries whereby science and technology in everyday life often disappear into the background. But, *given* the definitions used here, scientific and technological changes are *not* political, economic, and cultural changes; they belong to science and technology *only* if they consist of the advance of valid knowledge and ever more refined tools to intervene in and manipulate our environment.

The mesolevel view of science and technology as organizations—which is the main focus of this chapter—must therefore be supplemented both above and below: above, on the macrolevel, by changes in the increasing size of scientific output, the greater autonomy of the scientific community, and the greater scope of the sphere of science in relation to the other spheres of social life—and, as we shall see, its greater dependence on them in other respects. And below, in the workings of individual organizations promoting scientific and technological advance.

The Organization of Scientific Disciplines and the Relation Between Science and Nonscience

Where did the organizational strengths and the proliferation of institutions promoting science and technology come from? From the viewpoint of the organization of science, Whitley, following Collins (1975: esp. 490–92), stresses that the organizational autonomy of science is fairly recent, emerging with the university system in Europe in the late nineteenth century. Thus the argument of the first chapter about the distinctiveness of *modern* science and technology can be taken a step further: while it is possible to date the distinctive impact of modern science and technology at the latest from the time of the divergence in economic growth at some point during the early part of the nineteenth century, its present-day organizational forms did not take shape until after the closer coupling between science and technology, and with it the more systematic link to economic growth, during the second industrial revolution of the late nineteenth century.

With this autonomy, the institutional system as it has prevailed until today was preordained: "Academic norms and values directed research strategies and separated knowledge production from lay, amateur efforts so that truth became a monopoly, or nearly so, of academics" (Whitley 2000: 279). This has subsequently come to be the central importance of the university as an institution promoting scientific and technological advance: although industrial and publicly funded research outside the universities may have a much larger role in research and development (R&D), the cult of truth—to adopt the anthropological stance for a moment—is monopolized by academics, who are its priests and regarded as such by the lay-worshipping public.

The organizational autonomy of science includes the self-control of the scientific professions by means of the peer evaluation of output. This ability to police its own boundaries ensures not only the separate organizational base of the profession, but also its "relatively high degree of autonomy from the dominant culture, and from other professional groups" (Whitley 2000: 32). A second feature that now increasingly sets science and technology apart is "more complex and expensive" technical apparatus, "distancing the objects and procedures of science from everyday, lay concepts and substances" (Whitley 2000: 65). (This distance or "invisibility" of technical apparatus in everyday life in the form it takes in the research laboratory can be reconciled with the outward flow of laboratory technology into, for example, consumer products, as argued earlier in this chapter—the apparatus needs to take a different form before it can migrate into the world-at-large.) These social features—not epistemology—now separate science from other forms of knowledge production.

Jumping to more recent times, Whitley thinks that recent changes in the sciences have not been so dramatic as to warrant new labels like "the knowledge society" and the like. Instead, he argues that there are now increasing attempts to *manage* the process of knowledge production, and he points to more specific changes in the 1980s and 1990s: "State management of the public sciences has become more overt and direct in many countries . . . on the other hand, the end of the Cold War and reduction of military support of the physical sciences . . . together with the expansion and reorganization of the biomedical sciences, have reduced the dominance of physics as the icon of scientificity" (2000: xii). He adds expanding student numbers without increasing resources, plus social movements that have had an impact on the political support for science, as other sources of recent changes.

From this overall institutional autonomy of science, we can now move on to the autonomy of different areas of science—or disciplines within it— which are historically variable. Fuchs, following Whitley and Collins, argues that "the authority invested in scientific knowledge is ultimately due to the strength of the organizations that produce it" (1992: 192). More specifically, "fields with high levels of resource concentration and mutual dependence [of researchers on each others' results (my addition)] are more likely to produce facts than fields with low degrees of concentration and dependence" (1992: 89). This type of sociological finding is important for our understanding of the competition between scientific fields in any given period. I would add, however, that there is a larger context: the migration of prestige and resources from one scientific field to the next depending on where the research frontier is located.[4]

Whitley argues along similar lines that "the major differences between the sciences can be derived from . . . two distinct dimensions: *the degree of mutual dependence between researchers* in making competent and significant contributions and *the degree of task uncertainty* in producing and evaluating knowledge claims" (2000: 85, emphasis in the original). Task uncertainty in science operates in two opposite ways: "first, there is the institutional goal of reducing task uncertainty so that greater control can be exerted over the environment"—this is the side of advancing knowledge—"second, there is the professional need to maintain sufficient uncertainty . . . to avoid . . . external control of the research process" (2000: 139–40) on the side of the autonomy from society. Here we can already get a glimpse of the tension between applied and basic research in the R&D process translated into the context of the organization of the sciences. The larger context of all scientific disciplines, however, or of science as a whole, is that science has been able to manage this tension in relation to the public and so maintain a united front toward society.

Mutual dependence is the degree to which the disciplines or groups of researchers rely on each other's results. Whitley makes the interesting observation in this connection that the "most common form" of dependence of fields on each other stems from "the use of technical procedures and instruments from other fields" (2000: 268). Scientific fields thus become more alike in adopting common ways of working and similar organizational strategies, even if they simultaneously go in the direction of further specialization (2000: 271–76).

Task uncertainty and mutual dependence are only two variables—more can be added, and these two can be refined further. Furthermore, for different scientific disciplines, these variables change over time—physics is not the same discipline today as it was in its heyday in the mid-twentieth century. Nevertheless, physics is regarded as an example of low task uncertainty and high mutual dependence, with the social sciences at the other extreme. Chemistry and areas of computer science such as artificial intelligence lie somewhere in between, but the biosciences, for example, which are seen as the current leading edge of science, are hard to place since they have changed so much recently (and subspecialisms have proliferated).[5] In view of these changes and differences, we should not lose sight of what separates knowledge production from production of nonscientific knowledge in the first place: what sets knowledge apart from culture is not (just) epistemology, but that all the sciences share a common focus of attention, pushing knowledge forward. They have a common goal, a "sacred object" (Collins 1993: 202). A shared and open or "non-ideological" (Fuchs 2001: 7; see also Becher and Trowler 2001) network of communication sustains this common worship, but also enables scientific knowledge to keep moving forward.

There is no need here to go further into the variations among scientific disciplines on these two dimensions and how they have varied over time.[6] Suffice it to say that there is considerable variety in the strength of disciplinary organizations according to how highly integrated they are and how tightly defined their tasks are. It is easy to see, however, that the reputation of physics as the "ideal" of science rested on a high degree of integration and closely defined tasks, while the human sciences have been at the other extreme. The key is that the strength of the organization of a scientific field determines not only what sets it apart from other fields or disciplines, but also its advance relative to other fields: "The particular location of a field within that system," the system of "generating and co-ordinating intellectual innovations," Whitley concludes, "is a major factor in determining its autonomy, coherence, and direction" (2000: 266).

One difficulty therefore is not what sets disciplines apart from each other and what unifies science and sets it apart from nonscience (which has been much discussed), but how to trace the shifts in importance of different branches or areas of science and technology. For example, how could we say that one area of science and technology has become more important than another? Perhaps we could do so in relation to their diffusion and impact, but a moment's reflection makes clear that there is no immediate relation between "advance" and impact. This also explains why there is not *one* general reaction to science and technology (or to disenchantment), but ebbs and flows. Each wave, moreover, produces moral panics and cultural exuberance, or fears and enthusiasms. And, as Edgerton (2006) has pointed out, the constant focus on *new* technologies (and, I would add, science) makes for a very lopsided picture of their social implications. For a complete picture, it would be necessary to track the migration of scientific advance, and where science and technology are applied. Below, this will be done by selecting only *some* of the most sociologically significant advances and past patterns. A further limitation is that the sociology of science can only identify the strength or status of disciplines over the long run and with lots of hindsight—and lacks clarity about the strength or status at the current research front.

As for the strength of *different* sciences, it is necessary to counter an argument that has recently been made and that would rule out making comparisons between them. It has been argued that there is no such single thing as "science," or that there is a "disunity of the sciences" (Galison and Stump 1996). To distinguish different branches of science in order to trace their separate histories is no doubt useful, but the sciences also *share* important features: they are cumulative, unidirectional, and a boundary separates scientific knowledge from nonscientific knowledge. For knowledge apart from belief, the main divide is still between natural science and the humanities and social sciences, as argued by C. P. Snow ([1959] 1964). Collins's "rapid discovery science" is cumulative and moves from one area to the next; for the humanities and social sciences, in contrast, there is no rapid discovery because there is no "science-already-made" that can be taken for granted and thus no previously discovered territory to move away from—or previously undiscovered territory to move on to since the territory, the "objects" of the humanities and social sciences, are not as tightly delimited.

Here again we encounter "universality" and limits: openness in communication, as Fuchs (1996) argues, is a precondition for advance. He points to "standpoint epistemologies" (such as within certain variants of the sociology of science or deconstructivist literary theories) as obstacles to advance because they close off communication between disciplines by insisting that valid knowledge is bound by context. But there is also a wider implication here for the boundaries of scientific and other disciplines: these borders are where scientists (or engineers) cannot communicate with each other, but they are much less pronounced and often completely porous across the natural sciences by comparison with the social sciences and in the humanities, where there is often little communication or traffic between disciplines and subdisciplines.

It is also possible to make a broader comparison with culture: The reason why it does not make sense to talk about the unidirectional "advance" (or open communication within, or the unity of) culture is that, although with hindsight it may be possible to trace certain directions for culture, culture (high or popular) does not nowadays have a focus of attention (unlike in premodern societies, where legitimacy was attached to all-encompassing religions). Instead, culture is diffuse, comprising many elements not moving along a common front. A comparison with economic production is also instructive: like cultural change, economic change is plural, a pattern of growth in many different places. But like the advance of scientific knowledge and technological artifacts, it is constantly moving forward, toward accumulation. The difference is that they are cumulative in different senses; economic growth does not build stepwise on a common core.

Finally, an important pillar of the role of science and technology in society, apart from disciplinary organization, is the status of its professions. Fuchs summarizes the sociology of scientific professions as follows: "High task uncertainty combined with monopolistic expertise account for the ideologically elaborated mysteries surrounding professional work. Since lay clients and customers are usually not competent to judge professional work, practitioners are in a good position to credibly claim that their services are in the best interests of the laity" (1992: 147). And again, while Fuchs is right to argue that this reverend view of the profession should not be taken at face value, it is important to highlight the double edged nature of sociological analysis: we understand that there is a social basis for this prestige that should not be taken at face value, but we also understand the sociological "truth" that underlies it, which is that the separation between lay and scientific knowledge has a solid base in consumer products and other products of scientific and technological advance. (And we social scientists also rely on this prestige, even if our product, expertise, is usually not in demand in the same way.)

The status of scientific professionals (and technologists or engineers) is not just a matter of the prestige that the laity accords to knowledge. As Perkin points out, although advanced societies differ in the makeup and the strength of their professions, they share the feature that "as long as the particular expertise is scarce—science or medicine, administrative know-how or legal skill, computational dexterity or electronic engineering—the expert can command a rent as surely as the landlord or owner of industrial capital" (1996: 7). This is where the continually expanding knowledge in society meets "the market." Collins has identified the process of "credential inflation," whereby educational qualifications expand in number and in degree of specialization (Collins 1979). The advance of science and technology that has been described so far depends on this expanding and increasingly specialized stratum of knowledge professionals that, apart from the inwardly turned competition within the scientific community, has extended further and further into the wider society. Scientific knowledge increasingly dominates society, but the role of its practitioners in society at large and the status stratification within its own community are separate issues.

The Scientific Community and its Relation to Other Parts of Society

If we trace the autonomy of the scientific professions or the scientific community in relation to the external world, how it has become more well-established in modern society and recently become contested in certain respects, we can follow Collins in charting its changing battles with religion during the scientific revolution, its alliance with social reformers against conservatives in the nineteenth century, and jumping ahead to more recent times, the alliance with the military and economic establishment against dissident scientists and social movements such as environmentalists in the late twentieth century (1993: 308). But this needs to be put into a larger context than Collins, a conflict sociologist who sees competition among social groups everywhere (except on the local level of cohesion within the scientific cult), fails to acknowledge: the role of the sciences as a universal language or medium with a potential for creating and maintaining consensus above contending social groups. In the wider society, the possibility of maintaining this consensus rests on the unshakeability or lack of conflict over the belief that science provides an endless "bribery fund" of innovation.

It is true, as Collins argues (1993: 315), that in recent times scientists have been forced to ally themselves with elites and lay groups to finance their increasingly expensive research equipment. He also points out that the conventional view of science, that it automatically results in "applications" with economic benefits, is mistaken, and perhaps this is increasingly being recognized.

But here again we can see a limitation of a focus on conflict (again, Collins is a conflict sociologist) rather than cohesion: first, the stream of research equipment into everyday life is still going strong. Second, the belief that science leads in a straightforward way to practical benefits or applications, although it is mistaken in a blanket way, continues to be widely held among a lay public, even if it is increasingly being questioned by sceptical sociologists and other "insiders."

Again, Collins points out that big science needs the patronage of institutions with "deep pockets," unlike earlier science which could be financed by individual patrons. "Big science" (discussed in the next chapter) therefore means a more powerful role in society—but also greater dependence. This dependence, for Collins, is not just a question of resources, but also of allies: if the scientific community can avoid compromising the "purity" of its cult (or its research) with its institutional patrons, then it can maintain a unified front; if not, the community will split and will need to find other allies, for example, from among the lay public.

As we have seen, for Collins, the outward flow of research equipment into the everyday world enhances the power of scientists and engineers, but in this case lay interests also determine to what extent they do so. As will be argued in Chapter 6 on the consumption of technologies, here too Collins overestimates conflict. As Braun (1993: 25–26) has noticed, conflict over new (consumer) technologies is rather rare (we will come back to this in Chapter 6), and noticeable perhaps mainly precisely because it stands out so much. And apart from particular areas of conflict, the public does not determine the extent to which technologies come out of the lab insofar as they are continuously developed as if they will be accepted—and of course we generally become aware only of the successful ones. Further, we do not need to take the many "new" failures into account—since new successes are in any case always being generated, with—again—mainly social science experts noticing the extent of "failures."

Finally, Whitley points out that "the importance of being scientific may have grown in the twentieth century but what that means has changed and has been subject to alternative interpretations" (2000: 270). He notes that "science neither completely dominates the 'means of orientation' in industrialized countries, nor is it a fully integrated and unified organization with a single identity" (2000: 275). While agreeing with this, I want to press the point again that this view is limited. Whitley's view takes the competition between scientific fields as being a central feature of twentieth-century science; but it is equally important, in my view, to see the larger picture that "scientificity" has on the whole become more and more accepted as the standard for knowledge. The same is true of the organization of science: there is no monolithic "science," but scientific knowledge has a legitimacy beyond the content of individual scientific disciplines and institutions. We can put this point sociologically; scientists share a "cult of truth" or a "cult of accumulated discovery," and science as a whole also disenchants the social world and displaces the legitimacy of other forms of cognition—just as proliferating technologies add to and complement existing devices.

An important implication follows from this that has been overlooked: as mentioned earlier, the sociology of science has tried to identify the norms that bind scientists together in a community (Hess 1997: 52–58). Whatever these may be (and I have argued that they must at a minimum include Fuchs's commitment to open communication), I want to point to a feature of science that has not been noticed in the focus on binding norms; namely, that science also undermines social norms. It is possible to regard science as a "cult of truth," just like other belief-systems, but apart from reinforcing group solidarity within the scientific community, it also displaces other norms in society-at-large. For if the argument about science as disenchantment is correct, then science, by constantly extending the reach of certified valid (and in this sense value-free) knowledge, displaces other types of "knowledge."

To be sure, this displacement is not a zero-sum game, but it is to Weber's and Gellner's credit that they recognized this—otherwise strangely overlooked—effect (Fuchs and Collins, for example, overlook this). Here we can come back to "universality": disenchantment works because in modern society, as Weber put it, our belief-system rests on a "cosmos of natural causality" (1948: 355) in which "there are no mysterious incalculable forces" and "one can, in principle, master all things by calculation" (1948: 139). This applies to the relation between representations of the physical world and the physical world itself, and it is central to the definition of science here (and which this chapter took as its starting point): The Weberian "cosmos" is a belief that is oriented toward universality (*everything* is in principle knowable) and even though it does not exclude other beliefs in a zero-sum manner, it is the *potential* scope of this belief that matters here. Note the curious result of this view: scientists (and often engineers) argue that their work is value free, and my definition of and arguments about science support this view—*except that* the extension of value-free knowledge throughout the social world itself represents the strengthening and implementation of a value or a set of norms. Put differently, the domain of the "cult of truth" expands in society, leaving less room insofar as there is a zero-sum ground with culture (hence the science and technology triangle in Figure 1.1) that has been expanding at the expense of culture.

This also applies to the norms of scientific institutions. The sociology of science, again, has concentrated on how science polices the boundaries between itself and other nonscientific institutions (and sometimes also how disciplines within science draw these boundaries). Again, this emphasis misses one part of the larger picture: the important feature of science (as Fuchs also points out in relation to scientific communication) is that science is boundaryless within itself and toward other "knowledges" (scare quotes are needed only if science is considered to exhaust knowledge) and beliefs. In terms of the boundarylessness within science, we need only to think of advances where different scientific disciplines seamlessly meet each other. Science therefore erodes other nonscientific knowledges and norms, and the strength of science has generally been growing. Whether this process will continue indefinitely we do not know.

Finally, as Fuchs argues (1992: 17), the authority of scientific knowledge depends on the strength of its organizations. But this does not take into account the wider growing legitimacy of scientific knowledge and the prestige of more powerful technologies. Fuchs, unlike other sociologists of science, does not hesitate to identify advance and cumulation as a defining feature of scientific knowledge, and he specifies when this advance is likely to occur: "Under conditions of very high mutual dependence between scientists, competition among practitioners will lead to cumulation; i.e., to rapid advances in knowledge" (1992: 187). If we add these—organizational—conditions to the conditions that are "internal" to knowledge, "representing and intervening," and the technological conditions for advancing scientific knowledge via Collins's more powerful research instruments, we have the organizational bases for scientific advance.

Labels, Levels, and the Global Research Frontier

Before we turn to the wider role of science and technology in society, we must briefly complete the account of factors that contribute to "advance" but which also partly fall outside of it: R&D and innovation. In the case of R&D, the emphasis shifts from science to technology. The concept of innovation, on the other hand, is often used in economics, where the key question is how innovation promotes economic growth. The conventional definition of innovation is bringing a new product into the market or into practical use. Edquist defines innovations as "new creations of economic significance," which is broad enough to leave scope for some disagreements within the economics of innovation, such as whether diffusion should be included or not (1997: 1 and passim). The problem with using innovation in this way is that it includes not only scientific knowledge and technological artifacts, but also organizational change, diffusion, and other factors that promote growth outside of science and technology as defined here. Thus Edquist's definition departs from my definition of science and technology insofar as advance merely has to add value, and the notion of "advance," which is central to my definition of science and technology (intervening in nature, and manipulating the environment), disappears: innovations that add value may or may not be advances in this sense.

A further difficulty in obtaining a coherent picture of the various institutions promoting scientific and technological advance when we include R&D and innovation is that the different units of analysis and the concepts associated with them have not been brought systematically under one roof. Some of the units and concepts relate to individual areas of science and technology; big science in relation to particular endeavors such as physics, for example, or the large technological systems like electrical power. Some relate primarily to the national level, such as accounts of R&D spending and development, or public policy, or patterns of diffusion. Still others are both above and below the national level: for example, a single research group within a particular area, or the globalization of the scientific community (Drori et al. 2003). And finally, again, science and technology have become difficult to keep apart, which makes it even more difficult to separate levels and units.

Still, if we begin at the top, with global science and technology, we can recognize some phenomena that definitely belong to this level. On the part of science, as argued earlier (following Fuchs), science needs a commitment to ongoing open communication, and it is in this sense open to criticism (a "global" feature). Moreover, the density of communication networks at this level has increased, with an increase in the number of international scientific organizations and an increase in the number of publications (Drori et al. 2003: 4–6). The same applies to government support for R&D, for which Drori et al.

identify a global model with "regional variants" or "national styles" amidst an increasingly global model of national policies for pursuing science (2003: 196–213), even where they may not have the desired effect of economic development in the short term (2003: 221–48). And, as always when speaking about a global level, we need the caveat that global science and technology mainly means science and technology in developed societies, even if it is in principle global.

But we can also notice that at this level there is no way to bring the myriad of scientific and technological organizations under a single umbrella—apart from Fuchs's openness, and advance. The only systematic integration of these efforts is with concepts encountered earlier, such as postindustrial society or the knowledge society (Stehr 1994). These wholesale concepts, however, exaggerate a unified science because they take as *given* that the sphere of science or knowledge determines the other two—political and economic—spheres. Yet at the macrolevel, the distinguishing feature of R&D is that it consists of a decentralized and diffuse set of organizations—loosely tied laterally, vertically stratified by status, increasingly coupled to economic goals, and at the same time maintaining their autonomy in relation to the rest of society by means of a focus on advance and cumulation. It is also the case, as Collins has argued, that a limited attention space exists worldwide among researchers (1993) so that only the leading efforts in this advance will be at the center of attention at any given time. Still, a global logic of advance and communication and an intense stratification and competition are not mutually exclusive.

National Systems and the Changing Shape of R&D Organizations

When we come to the next level of individual states, one concept that has been prominent is *national* systems of innovation.[7] This concept has been developed mainly as a tool for policy making, though it clearly corresponds with a real trend of increasing efforts by governments to mobilize R&D for domestic economic gain.[8] As Edquist notes, "most public policies influencing the innovation system or the economy as a whole are still designed and implemented at the national level" (1997: 12). And we need to remind ourselves that a "proactive" government policy is a relatively recent phenomenon. To what extent these efforts have been successful need not concern us here since this a policy question rather than part of the more general issues addressed here. (Perhaps the key policy question is public funding for research. Mokyr says that "much of the history of technology in the twentieth century can be described as a continuous search for the right 'mix' of private and public efforts in R&D" [2002: 106]). Still, the concept of national systems of innovation is useful since it allows us to grapple with the fact that so much of the R&D effort, such as funding for research, operates at the national level.

One question that has recently come into focus is whether national systems of innovation are being undermined by an increasingly globalized economy? Here opinions diverge (see Pavitt and Patel 1999). For our purposes, it is enough to note that firms find niches for their innovations where they enjoy an advantage, and national or supranational monopolies may help them to do so. But viewed from a longer-term comparative-historical rather than a policy perspective, the new global competitiveness, which "makes easier the transmission of best-practice techniques across countries" (Archibugi, Howells, and Michie 1999: 11), is a process that has been ongoing in a deliberate way since before the first Industrial Revolution (Inkster 1991: 32–59), with the postwar period merely accelerating it (van der Wee 1987: 213; see also Chapter 4).

If there are many aspects to changing R&D institutions, the overall picture for the twentieth century is clear: rapid expansion in several waves over the course of the century, perhaps coming up against limits or plateauing in response to economic constraints in recent times. Scientific research and the development of new technologies are increasingly regarded as essential for economic advantage (and, though this is said more quietly, within the corridors of power and nowadays mostly out of earshot of the taxpayer, or within a specialist academic literature—for *military* advantage). Much has been written from a policy perspective about this, and we need only consider the aspects that are relevant to the argument here. The main question at this point (since the relation between advance and economic growth will be dealt with in Chapter 4) is: how has R&D changed apart from its connection with growth? We can summarize the main changes briefly: they range from big science as an institution (covered in the following chapter), the growth in output of publications and the number of professional scientists among the population (Drori et al. 2003: 4 and passim), the increasing density of communication among scientists (Becher and Trowler 2001), the reliance on increasingly complex and sophisticated apparatus (Collins 1993: 313–15), and the bureaucratic organization and differentiation of scientific specialisms. These organizations have, moreover, been subject to rationalization, just like other bureaucratic organizations in the economic and political spheres. At the same time, the centralization of organizations and unity of communication across sciences is compatible with differentiation or specialization and proliferation, since new disciplines and subdisciplines are constantly being added. Thus interdisciplinarity or mergers between disciplines do not represent dedifferentiation, but are complementary to differentiation.

A related point here is that the tightening of the relationship between science-technology and the economy produced a tension between "pure" and "applied" research from the start (Hounshell 1996: 16). This was already so for Edison's laboratory, which is regarded as the first American "industrial" R&D lab. Even at that time, some scientists questioned whether what Edison was doing could be included in science, and this tension has continued to the present day (though this also differs from one area to the next: pharmaceuticals have again become more basic or pure science, at least in the sense of being closely linked to university-based research [Rosenberg 2002]). Note also that this tension emerged at the time when it became difficult to distinguish between science and technology; for example, with science-based large technological systems. Finally, we can therefore mention areas *apart from* where R&D and large technological systems are linked: examples here are science-based industries such as chemicals and materials science—even though these are not directly tied to large technological systems, they nevertheless began to be produced on an industrial scale during this period, and in this way were tied to technology (for example, Freeman and Soete 1997: 85–136). At this point, we must therefore look beyond the organization of science (and technology) as such and turn to the joint emergence of big science, large technological systems, and the American system.

3 Big Science and Large Technological Systems

The Emergence of Big Science

In the twentieth century, big science and large technological systems were at the leading edge of "representing and intervening," "refining and manipulating," and "disenchantment." In the late nineteenth century, as we have seen, research took a leap in scale and organization. This development is described as a "Research Revolution" by Kargon, Leslie, and Schoenberger because, they say, it is:

> at this point that the notion of "research," especially in the natural sciences, began to transform society itself. First, science—especially physics and chemistry—became directly useful. Directed scientific endeavour—goal-directed research—became an *economic* factor to be reckoned with . . . Second, it was for the first time becoming understood that science could be an organized enterprise. If before the Research Revolution the production of science and scientists was (like lightning), an act of God, after it recognition was widespread that they could be produced like hats or pins. (1992: 337, emphasis in the original)

This concept of research has prevailed ever since, and this research revolution sets the stage for big science.

The research revolution began with the research lab, which first emerged in Germany in the late nineteenth century (Lenoir 1998)[1] and in the United States shortly thereafter, with another increase in scale along the way (Hounshell

1992). With the industrial labs of the interwar period, research also took on a large-scale organizational form. But the main impetus for big science came from the military and America's role in the World War II and later the cold war. Big science, however, is not the same thing as larger R&D labs. For this, we need not only the "directedness" of research, but yet another increase in scale: As Galison has pointed out, one defining feature of big science is simply size: "Big science is big relative not just to what scientists knew before, it is big relative to all science" (Galison, 1992: 2).[2] And Westwick says that for the national laboratories in the United States, "the scale of the labs themselves—their hundreds and thousands of scientists, their multi-million dollar budgets, their mammoth machines—dwarfed most other scientific institutions in post-war America, with the possible exception of industrial labs" (2003: 306).

Still, Hevly notes that even "big budgets and big instruments are only part of the story"; there are three other major features:

> the increasing concentration of resources into a decreasing number of research centers and the dedication of these special facilities to specific goals . . . second . . . the laboratory workforce has specialized [with] . . . hierarchies of group leaders, laboratory managers and business coordinators . . . third, big science . . . depends on the attachment of social and political significance to scientific projects, whether for their contribution to national health, military power, industrial potential, or prestige. (1992: 356–57)

These features set big science—where, again, the leading edge of science was located in the second half of the twentieth century—apart from other forms of science.

The two most well-known examples of big science are perhaps the first and the most recent: the physics of the atomic bomb, and the human genome project. Both illustrate that science and technology are almost impossible to separate since both involve the use of purpose-built technologies—for manipulating particles in the first case and using automated instruments for sequencing genes in the second. Both also illustrate that it is difficult to separate big science from the rest of society. Galison's description of particle physics is a good example: "In a collaboration of 800 [physicists] on a billion-dollar device . . . the experimenter could no longer select or quickly modify the goals of an experiment . . . experimentation with 800 colleagues is itself an unproven social and epistemic experiment" (1997: 44). The same could be said for the human genome project, with its race between different international groups and competition between publicly and commercially funded efforts.

In other words, big science means orienting a complex organization and expensive pieces of machinery toward a particular scientific goal. There are some direct social implications. As Galison puts it, "by its very size, big science cannot survive in isolation from the nonscientific spheres in society" (1992: 17). Collins is more specific: "the technologies that scientists use in their research require huge economic inputs, and research technologies themselves cross the border into the lay world on an increasingly large scale . . . in order to carry out the struggle for Truth at the research frontier, scientists need to make an unholy alliance with mundane interests. The ramifications of this need are at the center of most current science/society conflicts" (1993: 309).

Apart from scale and increased resources, one feature that is easily over-looked is that big science is driven—in the economy, in public research, and in the military—by the notion that, if enough resources are directed toward a problem (or "enough money is thrown" at it), it can be solved. Edison is the starting point for this way of doing research, tinkering with apparatus in a systematic way. As pointed out earlier, this is different from how science worked before the twentieth century, but the harnessing of science on a large scale for specific goals also sets it apart from other forms of science, which makes for an especially acute tension between "pure" and "applied" science because it immediately raises the question of the aim to which vast resources are devoted.

The pure scale of organization also produces complexity: Whitley suggests that " 'big science' is likely increasingly to encourage concentration of control in science and thus increase the degree of mutual dependence and co-ordination problems" (2000: 108). In other words, big science faces the problems of all bureaucratic behemoths: steerability and inertia. But the close coupling of many elements in big science efforts also sets it apart from other forms of large-scale research: big science at the research front is more tightly organized—or centralized—than other areas of research. Centralization does not exclude a small number of large sites pushing on related parts of a common research front—these may nevertheless constitute a "system." As Westwick says of the postwar national labs in America, "competition stemmed from the organization of the national laboratories as a *system* of several labs, as opposed to a single central facility" (2003: 1, emphasis in the original, see also p. 312).[3] At the same time, this tight interdependence is a feature that big science shares with large technological systems, to which we will turn in a moment.

A more recent novel feature is that the organization of big science has become international. To put it in network terms: "In big sciences requiring a lot of expensive and sophisticated hardware, a major part of the work is securing new instruments, linking them together in complex chains and networks, and tinkering with them to detect more direct and straightforward traces of novel phenomena. Such organizations become centers anchoring extensive international networks of cooperation and competition" (Fuchs 2001: 59). Since this feature sets it apart from large technological systems[4] and from "national systems of innovation," it is worth stressing: it is inconceivable now—though this was not the case for early big science—that a big science effort should be initiated without taking international competition and collaboration into account. Big science as a way of organizing science has thus become an important part of the links between the major developed societies.

Large Technological Systems

Hughes's (1987) account of the evolution of large technological systems (a term he coined) is well-known, so I will concentrate here instead on key features that relate to my argument. To do so, we can borrow a succinct definition of large technological systems from Joerges, who says they are "ensembles of artefactual technical structures and their non-artefactual technical components which are (a) integrated (coupled, networked) across extensive spatial and temporal reaches and (b) enable and ensure the functioning of a large number of other technical systems and thereby (c) link their organizations together" (Joerges, 1992: 56; cf. Hughes 1987: 51). Before the late nineteenth century, there were few technologies that could be called large technological systems. Waste and water networks, for example, covered large areas with interdependent parts in ancient Rome and earlier, but they do not meet Joerges' conditions (b) and (c).[5] In any case, in the late nineteenth century, several large technological systems emerged, and these are still the ones we associate with the concept. The period of rapid industrialization, 1870 to 1920, was the high tide of system building in America, one result of which was that, as Cowan puts it, "people become less dependent on nature and more dependent on each other" (1997: 149).

Large technological systems can thus be seen as the technological infrastructures of modern societies. Energy, transportation, and communication are the main ones we typically think of, but waste and water are examples of less visible ones. These are also the major ones in developed societies, though each of these can be further subdivided into subsystems; for example, transportation consists of road transport, rail, shipping, and so forth. We should also remember that not all parts of society's infrastructure are technologies: education, for example. And not all large technological systems are part of society's infrastructure: military large technological systems are an example. Still, Kajser (1994) uses the term *infrastructual systems* for those large technological systems that were created during the period when the state began to pentrate more deeply into society (Mann uses the term *infrastructural power* (1993a: 59–61) in connection with enhanced state capacity in this period). Here again it can be seen that technology—just like science—can be kept analytically separate from society, even if it is also entwined with patterns of social development.

Before we go into some of the implications of these systems, some further characteristics can be highlighted First, large technological systems always bring together technological artifacts and social institutions.[6] Second, these systems have an inbuilt momentum or trajectory. As Hughes puts it, "as they grow larger and more complex, systems tend to be more shaping of society and less shaped by it" (1994: 114). Third, this trajectory means that systems become "locked in." The second and third feature entail that even though large technological systems may start with some room for choice, they will become increasingly rigid. Fourth, the concept is functionalist, which entails that it must be embedded in an environment that has a "need" for this increasing intertwining and increasing scale and scope.

This last characteristic is worth examining in more detail. In the United States, these systems and the rise of modern business systems coevolved. Hounshell has noticed the similarities between the growth of large technological systems described by Hughes and Chandler's account of the organizational development of the modern multidivisional firm: "Technology appears to be the key ingredient in Chandler's take on bureaucratic rationalization" (1995: 210).[7] For Hughes and Chandler, how large technological systems and multidivisional firms extend their scale and scope follows a functionalist logic. Chandler's account of the modern industrial enterprise has often been criticized among economic historians (for example, Perrow 2002: 229–30), but my use of his ideas here relates not to his argument about the economic superiority of this form of the firm, but how it relates to technological advance in the sense used here. The point is that this new firm, because of the impact of technological advance, was more powerful: Chandler argues that these new firms "no longer competed on the basis of price: Instead, they competed for market share and profits through functional and strategic effectiveness" (1990: 8), which, in turn, critically depended on technology. The evolution of the American firm can thus be seen as a combination of Hughesian functionalism with my notion of technology as "refinement." Again, the debate about Chandler continues in economic or business history about whether his account is correct in terms of explaining the development of American capitalism; his idea that firms under capitalism develop toward greater efficiency has been challenged by others who put greater emphasis on ownership issues (Perrow 2002; Roy 1990). This is not important here; what *is* critical is how a process of "rationalization" is central to economic development (we will come back to this shortly).

But first, we should note the limits of this functionalist logic: a large technological system tries to bring its environment under control; once it has done this, it is a "closed" system. Moreover, big science and large technological systems have transformed the natural and social worlds *together*. As Hughes points out, the transformation of scientific research and large technological systems are linked: "A major reason for the gradual displacement of the independents who had modelled their behaviour on the young Edison was the need of corporations like Bell and General Electric to preside over the expansion of existing technological systems. To this end, they wished to choose the problems that inventors would solve, problems pertaining to the patents, machines, processes, and products in which the corporations had heavily invested" (1989: 151).

A further feature of this functionalist logic is that the human environment has been transformed: "Since 1870 inventors, scientists, and system builders have been engaged in creating the technological systems of the modern world. Today, most of the industrial world lives in a made environment structured by these systems, not in the natural environment of past centuries" (Hughes 1989: 184). This control over the environment lies partly at least outside of the economic realm or even outside the human environment, affecting not only politics and militarism and our everyday way of life (or culture), but also the (extrasocial) physical environment. It is worth quoting McNeill again (this was quoted in the introductory chapter) in the context of big science and large technological systems, who calls our transformation of the environment "a gigantic uncontrolled experiment . . . for the most part, the ecological peculiarity of the twentieth century is a matter of scale and intensity" (McNeill 2000: 4).

The links between large technological systems and the economy go further. Thomas Edison was, in Hughes's view, the first "system-builder," and his main argument is that Edison and those like him inevitably underwent a transformation from inventors into entrepreneurs as their systems became larger and more complex. System builders increasingly needed to mobilize social resources at the same time as making technological improvements as their systems extended their scope and control over the environment. Another good example, also described by Hughes, is Walther Rathenau, a "financier 'system builder'" (1990: 11) who tried to create the appropriate large-scale organizational setting for promoting technological innovation in, among other things, the German energy system. In Rathenau's case this meant rationalizing, or in his own terminology "mechanizing," "the organization of work, transportation and communication, research, capital flow and political economy" (Hughes 1990: 13). Edison and Rathenau therefore also, to look ahead to the chapter on economic change, provide the link between the economy and innovation: entrepreneurs who, at least for part of their careers, were able to combine invention with the ability to generate the financial resources to exploit the available market opportunities (as Schumpeter argued).

But there is another way of thinking about large technological systems apart from as innovation—since these systems also become "frozen." Thus Nye (1997: 1076, 1086) argues that large technological systems—his examples are communication networks, but the same could said, for example, for energy systems—have always been monopolistic or oligopolistic in the United States. Economists conceive of this as "rational" because "each additional user" brings extra profits that outweigh costs, an effect they call "positive" or "network externalities." But this is an odd way to think about large technological systems: from the point of view of systems developers, it is equally appropriate to conceptualize their position in terms of market dominance (or monopoly), and from the user's perspective it is appropriate to consider both the benefits and drawbacks of this dominance. The competition among different providers of digital content or different software standards offers some more recent examples.

With this limitation, we come to the context in which these systems are embedded: large technological systems can be described as networks, spreading outwards and covering—or fanning out into and filling, often until they reach national borders—all available niches. But when they have done this and been institutionalized and technically standardized, these systems are difficult to change, a feature that is often mentioned—without, to my knowledge, drawing out the full implications. This leads us, finally, to how we should see these systems among wider patterns of social change: large technological systems, like big science, are new; they are unlike what went before. But unlike big science, which migrates from one field to the next and advances cumulatively, and although these systems can be added to, once in place, they are mainly refined or their capacity changes only incrementally, and wholesale changes become virtually impossible. They are "in place" rather than migrating, unless they are slowly replaced in the sense of being superseded when new advances come along—the telegraph is a good example. And unlike (some) stand-alone technologies, they have become indispensable to the functioning of social life, and this "functional" nature makes them into the part of science and technology that is most deeply embedded in our everyday lives and our environment—at once pervasive and at the same time "invisible." Here we can think of many examples, including the apparatuses relying on electricity or the vehicles that transport us daily.

The concept of large technological systems thus allows us to grasp the development of artifacts, from invention to routine use, as part of an ensemble of interconnected technological and social parts. Second, the concept allows us to grasp the powerful role of this type of technology in society: large technological systems bring large parts of the natural and social environment under control. But large technological systems can also be seen, apart from a functionalist evolutionary perspective, as bureaucratic organizations combined with technologies to which we are subject or which organize people in particular ways. And although, as Hughes argues, these systems are initially fluid, once they are in place their logic is a soft determinism (1994). There may be incremental changes, but there will also be an inbuilt stagnation.

The appropriateness of Weber's notion of "caging" is therefore obvious here since the conservatism of these systems locks us in even when improvement (refining and manipulating) are possible. Weber spoke of "routinization" (or "ossification" or "petrification") and the "iron cage," and of technology as "frozen spirit" (1980: 322, the English translation is sometimes rendered as "congealed spirit," 1994: 158). Rathenau similarly worried about the advancing rationalization of research and of large technological systems: the "mechanization" of the world, as he called it, would "strangle the spirit in a network of organizations" (cited by Hughes 1990: 26). But the notion of an exoskeleton is equally appropriate in view of what was said about living in environments that are no longer natural but human made: these are the most powerful technological exoskeletons that we have, and even if their enormity constrains us in many ways, they enable us in many others.

Scale, Scope, and Limits

Big science and large technological systems are where the tentacles of scientific and technological development *as such* reach out furthest into the world, where there are the most systematic attempts to intervene in and manipulate the physical and human environments. Hence the point about the separateness of the two from the rest of society deserves to be made again: many within the tradition of the sociology of science and technology have argued that the interpenetration of big science and technology with society means that science and technology *are* cultural, or political, or economic—they are indistinguishable from society.[8] And, as mentioned earlier, it is true that big science projects need to mobilize societal resources and that large technological systems are more entwined with society than other technologies. The regulation, maintenance, and standards of these systems, for example, entail that technological issues and social ones are hard to separate. But this does not mean that analytical distinctions have to be abandoned. My argument is that while big science and large technological systems are interrelated with society, this is not what is essential here, which is that these two institutions transform the extrasocial world, nature and the environment—and *with it* the social world—and that they do this separately from other political, economic, and cultural institutions. Thus they create a new, continuously expanding social cage, or again, a massive exoskeleton. And the point about *big* science and *large* technological systems is that there has been qualitative change in this respect—and the distinct labels of size for these late nineteenth- and twentieth-century phenomena are therefore deserved.

Since this is where science and technology stretch furthest outwards, it is useful to identify their boundaries or limits more closely: One is the ceiling of R&D spending and the areas of big science that are in an expansive phase as opposed to a stagnating or declining one. Another is the slow incremental growth in the number of large technological systems and the fact that these systems (and big science too) are subject to political regulation and steering.

These are therefore also the broader limits or boundaries: new large technological systems are only gradually being added, and big science efforts migrate slowly, with both subject to societal constraints.

"Speculative" determinists can thus be criticized from this perspective as much as constructivists: the limits of big science and large technological systems mean that claims that science and technology have limitless possibilities are sociologically unrealistic, and this is borne out by the slowness and limited range within which big science migrates and by the conservative momentum of large technological systems. This point can be made differently: with the difference in scale in the twentieth century, there had at the same time been a qualitative shift in the types of problems that were addressed by science and technology (examples that come to mind here include the "war" on cancer and space exploration). But we can also discern the horizon, the limits of big science and large technological systems: it may be that we cannot gauge the size of the problems that can be solved in the future, or what is scientifically and technologically feasible, but what we *do* know is that there are limits—economic, political, and perhaps also cultural—to the resources and energies that can be directed at these problems.

At this point we can come back to the very definitions of science and technology and to questions about social structure and social change. On the side of technology, these are the deepest and widest reaches of refining and manipulating and also the place where this process has become most frozen and where we can therefore see some of its limits. It is similar with representing and intervening, where this process is constrained by the resources that are concentrated at the research frontier. Thus the definitions presented here can be related to structure and change: The first point is that this is an ongoing process and thus cannot be captured by the normal parts/whole distinction or structural conflict theories. For example, large technological systems have a functionalist (developmental) logic, but they have also become frozen. Second, both institutions changed in the twentieth century, in scale and scope, and in this sense they settled into the overall social structure, becoming large, indivisible, and simultaneously (and paradoxically) invisible parts of it. And finally, this is a (now large) part of the social (infra-) structure and an inescapable part of modernization, and its limits and possibilities are shaped by this larger process. Hence the problem that neither a conflict nor a functionalist perspective can capture these two phenomena in their entirety.

We can also at this stage draw out another implication of the argument, which is about the diffusion of these systems. As Inkster points out, technology transfer did not play a great role before industrialization (1991a: 298). During industrialization, of course, there were intensive efforts to emulate the most rapidly growing countries (Weiss and Hobson 1995). Nowadays, however, technology transfer and diffusion are fundamentally different even from their role during industrialization, "for two major reasons" according to Inkster:

> First, technology is fundamentally different: twentieth century technique embodies a far greater scale, capital, knowledge and skill content than that of the eighteenth century. Second, the mechanisms whereby technology is transferred from one locus to another have also changed radically. The role of individuals, of basic, generalized skills, of trade and of markets seem far less than the role of transnational corporations, of highly specific, scientific knowledge and of nonmarket contractual obligations. (1991a: 301)

In short, to put it in terms of the argument advanced here, technology transfer is more about systems than about individuals and individual organizations, and this aspect of science and technology has not yet been sufficiently recognized (see Chang 2002: 55–57).

Big science and large technological systems also allow us to reflect on the role of science and technology in society as a whole in a different way: while the effects of science and technology in everyday life are so widespread that they often escape our notice, big science and large technological systems are more visible in a different sense; there is a single migration at the research frontier where attention is focused, and there has been a small number of large technological systems that have settled into an ossified form and on which we depend constantly—such that they are highly visible when they fail or fall short. This visibility means that we can trace the leading edge of science (but cannot see where it will migrate next), and we can follow the development of large technological systems (but again, cannot look beyond them to their replacements or demise).

Entwining, Momentum, and Frozenness: On the Shoulders of Behemoths

Big science and large technological systems "entwine" with each other, and with society. *Entwine* is a term used by Mann to indicate power sources whose "interactions change one another's inner shapes as well as their outward trajectories" (1993a: 2). The entwining of big science and large technological systems varies and would need to be specified for each case—but in all of them, technologies have to be made more powerful to advance science, and science has to become more powerful to improve the workings of technology. An obvious example is how computer science and technology are interlinked in the advances of today's communication systems. Both together change the possibilities for manipulating the environment and extending the "footprint" of the mastered environment.

Yet big science and large technological systems are also entwined with *society,* foremost in their mode of organization. They have become increasingly bureaucratized, and rationalization has made them into organizational behemoths. We can outline the main features of this process following Weber, for whom "bureaucratic administration means fundamentally domination through knowledge. This is the feature which makes it specifically rational" (1978: 225). Knowledge in Weber's sense is broader than science or technology. Nevertheless, as we saw earlier for Edison (and even more self-consciously so for Rathenau), rationalization and bureaucratization invariably attend the development of large technological systems—and big science. And although I have kept knowledge and science separate so far, we should note a further entwining, via the increasing scientificity of organizational knowledge. Thus Dandeker describes an important part of the transformation of the state and the commercial enterprise as consisting of "the increasing knowledgeability of organizations . . . the development of increasingly elaborate and intensive systems of collecting, storing and processing information about the internal and external conditions of organizations" (1990: 197). Dandeker therefore puts more emphasis on caging than Chandler, who emphasizes greater effectiveness. The difference is clear when Dandeker says: "No matter how much the technical and organizational attributes of industrialism can be shown to have been originally the products of the capitalist enterprise system, they are relatively autonomous from it: they can be adapted for use in quite different institutional contexts" (Dandeker 1990: 204). Or again, "the rationality of bureaucratic discipline was an outcome of the technical superiority of this type of administrative system as an implement in the eternal struggle for power in social life" (1990: 206)—in other words, enhanced power, not economic efficiency, is a key to this process.[9]

Large technological systems, with their enhanced organizational capability, are therefore a major building block of the scientific and technological cage we live in. They are social cages in the sense that the enhanced mastery over the natural and social environments entails that these systems constrain (often literally) as well as enable many aspects of our social lives, including often "invisible" costs to the environment. (And these costs may themselves impose further constraints—as well as require the system to be extended still further). They are also invisible cages, inasmuch as they are visible mainly when they are first created, often inspiring awe. Nowadays, even their ongoing refinements are invisible, although a major retooling of large technological systems (as when a new technology is introduced into parts of the system) may still cause a stir. Otherwise, these large technological systems have a "momentum" (Hughes 1994) of their own, or not so much a momentum as an inbuilt stagnation. And even though large technological systems are not immutable, their systemic character means that a number of different factors have to come together in order to change them (see Summerton 1994). This is not to say that large technological systems are completely "frozen"; they change especially in the process of attempts to expand them or in response to changing conditions in their environments (political regulation, new competing systems, changing consumer habits, and so forth). The point remains that these systems are very stable, and their impact is therefore also a "given" feature of today's society.

There has been much discussion of how complex these systems have become, and how "normal accidents" or breakdowns in "tightly coupled" systems can occur (Perrow 1984). But again, what is even more striking is how stable these systems are. Conflict, if there is any, is mainly settled by compromise once the system is in place. It can also be mentioned here that there is little resistance to large technological systems—nuclear power is "the first time" that "a major online technology has been successfully challenged" (Rucht 1995: 277), and success in this case is still questionable.[10]

The cost of large technological systems is fairly fixed or changes only incrementally, and likewise the political economy of these systems—we can think of how the much-trumpeted liberalization of various infrastructural services such as energy in the 1980s and 1990s has fizzled out and done little to change the nature of these systems. Technological change in these systems is mainly routine—the equivalent of normal science or science already made. And in view of the argument made here about the environmental footprint of these systems, there may be advantages to this frozenness: the extent to which they continue to transform the environment is limited by how frozen they have become, unlike big science, where the leading edge continues to migrate, even if big science efforts are also bureaucratic behemoths with a conservative momentum.

The most important debates about the future of science and technology thus also take place in relation to big science and large technological systems, because the most critical policy issues for the future have to do with where our footprint is largest in the environment, and hence with the limits and possibilities for directing this impact. The argument put forward here, especially about large technological systems, suggests that they have a conservative momentum, and they are largely frozen. This might imply a pessimistic outlook, and it is true that the large technological systems that are in need of most urgent modifications are difficult to change because of the interconnectedness of their parts. Moreover, the creation of new large technological systems, for energy, for example, is possible, but many parts of the system would need to come together fluidly, and we are still at an early stage of knowing how early technology choices for large technological systems can lead to the successful creation of effective and sustainable uses of technologies. Something similar could be said here about big science, if we consider the impact, for example, of how current advances in biotechnology may shape the future and the unforeseeability of the consequences of early choices. Put differently, how to create viable "package deals," or how to steer these juggernauts, is one of the main challenges for the future of large technological systems and big science.

Some Swedish-American Comparisons

The focus so far has been on the United States, so a brief comparison can be made with Sweden, which will be taken further in later chapters. The point of comparing large technological systems is to show that, although the course of their development is somewhat different, the result is to all intents and purposes the same.[11]

Kajser argues that the development phase of Swedish infrastructural systems departed most from international patterns for the rail, electricity, and telephone systems (1994: 99). What accounts for this difference? For Kajser, the difference between Sweden and the United States lies in the balance between the central state and regional development on the one hand, and between public and commercial development on the other. "The 'Swedish model' in infrastructure system development," he says, "was that the state had the responsibility to develop the national 'grid' while responsibility for the regional network or components were left to local government and private enterprise," plus "informal cooperation between these actors and an avoidance of state controlling agencies" (Kajser 1994: 179). Thus the state's involvement was mainly limited to technical standardization and aimed at rapid economic development (1994: 179). This contrasts with the United States, where "private firms were allowed to build up extensive regional and even national monopolies" (1994: 183).

Yet even at the outset, when we look at the emergence of the different large technological systems from a comparative perspective, America and Sweden look quite similar. The key phase of development took place in the late nineteenth and early twentieth centuries, and even though Sweden industrialized somewhat later and its systems lagged behind those in the United States, in 1885 Stockholm had more telephones (5,000) than any other city in the world (Kajser 1994: 115). And the similarities—in transport, energy, communication, waste, and water—are striking: the supportive and regulative role of the state, the expansion from a core to a national grid, and above all the increasing dependence on a number of such systems (Kajser 1999 for Sweden; Cowan 1997: 149–72 for the United States).

Kajser argues that the most distinctive feature of what he calls the "institutional regime for infrastructural systems" (or here, large technological systems, minus those that are not infrastructural) in Sweden is the "helping hand" of the state.[12] But even though he contrasts Sweden with the more market-based institutional regimes in the United States and Britain, he also notices that recent moves toward liberalization or the encouragement of private sector competition have unsettled the traditional and stable Swedish regime and perhaps brought it more closely in line with other regimes (Kajser 1999). Further, he points out that the main current challenge for infrastructure systems are environmental ones, which is of course a challenge that is not faced by Sweden alone.

It is worth noting that the differences between the institutional regimes and their systems that remain are political rather than cultural. The (political) place where much of the entwining between large technological systems and society is located is of course the state. Here we would therefore expect to find the largest differences between large technological systems in the case of two countries that are as far apart as they can be in the comparative political sociology of Western developed societies. What we find is that, despite different political entwinings, the infrastructural large technological systems have become more similar over time: to be sure, we can make a list of some of the major differences—more decentralized (U.S. federalism) and market driven on one side, and more governed by a centralized state and in accordance with a social-democratic economy on the other. But again, several important similarities are equally striking: nation-state boundedness, the capabilities of (and problems faced by) the systems, their reach and scope, and (as we shall see in Chapters 5 and 6) usage patterns—for all of these the similarities outweigh the differences. This is even more true on the phenomenological level: Swedes and Americans would, I wager, be hard pressed to describe the differences between their respective infrastructural large technological systems (with the possible exception of transport, which we will come back to). Put differently, from the point of view of the user, or the consumption of the systems, the differences are invisible.

The point can be widened. A number of studies have traced the social factors shaping individual large technological systems (for an overview, see Summerton 1994). What has been much less commented upon is how they have become ever more similar. Hughes (1983), for example, has discussed esthetic differences—"styles"—of different energy systems. But this reinforces the point I am making, since underneath this stylistic veneer, the systems are (especially technically and socially) very similar. Take the well-known example of different systems of alternating and direct current (AC/DC) electricity systems. One can ask, what is more important, this difference or the overwhelming similarities *in effect* between the two systems? In other words, even when standards are different, the effect—powering appliances—is the same. Certainly the end result appears very similar to the user: we use these systems in similar ways, and we don't notice the systems in any case. The exceptions are mainly when there are political attempts to change the system, but this tends to be when they are new or new elements are introduced into them.

In certain respects, the differences are important. Energy consumption by Americans is approximately twice that of Europeans per capita (Nye 1998: 223), and the transport system presents the users with different possibilities and constraints. But the systems with their conservative momentum are now largely "locked in," and so I disagree with Nye (1998: 1) when he argues, in relation to energy, that it is our way of life or culture that matters most, with the implication that cultural change (or "reinvention") can alter the trajectories of the systems. My argument suggests instead that mainly even greater technological mastery can cope with the problems generated by our increasing mastery of the environment via large technological systems. The implication is that the challenge for economists, politicians, engineers (and users, by means of their uses—not culture) is to jointly improve the systems.[13]

Another implication of this comparison relates to the point made earlier about technology transfer. The differences between systems bear most importantly on the question: how does the development of the whole set of large technological systems, the technological infrastructure of society, affect the role of social or "industrial development" (again, Inkster's [1991a] broader term, instead of economic growth) since, as we have seen, they are indispensable to modernization? This is a question of timing and effectiveness. Perhaps research should focus on this question, in view of its obvious relevance for developing societies, instead of the different styles of the systems. And again, within developed societies, perhaps the main question concerns the extent to which frozen systems can be unthawed. Otherwise, and once they are in place, again, these systems are rather similar.

A Note on Science, Technology, and Militarism

Big science and large technological systems are closely associated with militarism so that, even if it cannot be covered in detail due to a lack of space and a lack of fit with the other topics here, some brief comments may be useful. Militarism would require a lengthy analysis of geopolitics, but for our purposes we can follow Mann, who has identified the key features which set the cold war apart from previous geopolitical periods. He argues that for the two superpowers in the post-World War II era, science was a key component of "deterrence-science militarism." During the cold war era, he says, military technology changed, increasing "privacy . . . in three ways": "the technology is more separate from the rest of the economy" than it was, it has become "more abstruse and scientistic," and there is "secrecy about weapon development," which is at odds with open democratic government (1988: 178–80). Note that these features apply to both former superpowers and to others with nuclear weapons, and that this privacy and secrecy are part of *political* or part of military power,[14] and not intrinsic to the content of science or the process of refining the technology.

Much has been written about how the cold war influenced science and technology, and one main line of thinking here has been the "distortionist" view, or the idea that scientific knowledge and technology were subverted to military ends. So, for example, Leslie has described how the U.S. military shaped the research agenda in some of the foremost American university laboratories, at MIT and Stanford, peaking during the 1950s and 1980s (1993: 2). But it is important to consider closely the claims being made for the distinctive features of science and technology in the military realm. To be sure, there were areas of science and technology that were made subservient to military aims, and science and technology enhanced the capability to pursue antidemocratic and dangerous political strategies. Mann's notion "deterrence-science militarism" includes the argument that science and technology are beyond the democratic control of society. But again, this did not produce distortions that were *intrinsic* to the pursuit of science and technology. The communication of results was sometimes "closed" or separated from the rest of the scientific community, and the "driver" of science and technology was not (directly) economic growth in this case; resources were allocated to military purposes rather than industrial development. Yet this is not, *pace* MacKenzie (1992) and Edwards (1996), an argument for the sociocultural determination (social shaping or constructivism) of science and technology since the content—the results—of science and technology were not distorted. One simple indication of this is that the features of deterrence-science militarism apply to both superpowers—and perhaps beyond the cold war to several nuclear powers (Mann 2001).

Outside the cold war period, Collins (1986: 85–92, 167–85) has argued that there is a more general link between the leading edge of geopolitical power and technological advance, but interestingly he suggests that during the cold war this applies to air power rather than to nuclear weapons. And again, there is no intrinsic connection between *American* science and technology and military power since Collins argues that this type of geopolitical-cum-technological advantage applies to all leading powers:[15] the link is specifically between science and technology and geopolitical advantage, and it is contained within this area rather than being a link whereby a militaristic society as a whole drives scientific and technological change in a particular direction, or science and technology are by their very nature harnessed to military goals. Perhaps, as Hounshell (2001) suggests, it is now possible to ask, after the end of the cold war, whether the focus on the connection between militarism and science and technology was itself a temporary distortion in the study of science and technology that drew attention away from more enduring issues. Put differently, although specifically military areas of scientific and technological advance will no doubt continue, they too can, after the end of the cold war, be put into a "contained" context, rather than being *the* defining feature of scientific and technological advance as they were within certain areas during the cold war interregnum.

The tightening relationship between science-technology and the economic sphere was the most important change for science and technology in the twentieth century (as we shall see in the next chapter). If, during the cold war, the powerhouses of research were oriented to military advances, it is not clear what sociological analysis can gain from this except a "what if . . ." counterfactual question—what if research had been used for civilian purposes instead?—which is bound to be speculative or inconclusive. Apart from this, most militarism is removed from our daily concerns, so that we, in the "Northern zone of peace," can *now* also leave the consequences of military science and technology for our everyday life out of the discussion (if we are lucky, we may never know). The impact of the militarism of developed societies on the developing world has, in the meantime, become ever more devastating (Mann 2001).

Nuclear weapons are clearly organized as big science and large technological systems. Unlike other large technological systems that are "functional" and part of the national social infrastructure, military large technological systems go beyond functional needs and extend beyond national borders. The Swedish and American military "large technological systems," for example, can be seen as supranational and to some extent dependent on each other. And at first sight, they also seem to be more politically shaped than other systems. Thus Weinberger (2001) argues that the Swedish dependence on American military technology during the height of the cold war and the need to coordinate its defense with the United States and NATO pushed Sweden away from its official neutrality policy and towards an implicit alignment with the U.S.-led alliance. This seems to be a good example then of technological systems spanning across borders. The interesting consequence that Weinberger draws from this example, however, is that large technological systems—in this case, air defense—have the power to shape other social forces—here, the geopolitical alliance between the two countries.[16] Militarism and its technologies fall outside the scope of this study, but it seems worth mentioning an example from the two main cases used here whereby even this most political of technologies can shape—as well as be shaped by—the relations between states.

4   Science, Technology, and Economic Change

The Great Divergence, the Invention of Permanent Economic Growth, and the Routinization of Progress

The argument made briefly in Chapter 1 linking the modernity of science, technology, and economic growth can now be taken further in this chapter. But first, we need to step back briefly from this topic to reflect upon the extent to which we have come to take this link for granted. For policy makers and for the public at large, innovation and economic progress (or growth) are often synonymous. In Inkster's (1991b) view, the equation of R&D policy with economic policy has recently been strengthened, with academic analysis and political concerns reinforcing each other in the competition between advanced societies. But, as should be clear by now, we can hardly take the ideas of those who promote science and technology at face value, nor can we rely on popular perceptions of technology and economic change. Instead, we need a longer term perspective that specifies the mechanisms of this relationship, including how it is conceptualized.

The very idea of a separate process for the invention and development of technologies for commercial gain was itself, as we saw at the beginning of Chapter 3, a recent invention.[1] This chapter will not go into technical debates in economics about the relation between technology and productivity.[2] It will suffice at this point to mention that common perceptions of this link leave out the contribution of militarism to economic development (Dandeker 1990) and that there is element of myth around the contribution of technical expertise to economic growth, the "myth of technocracy" (Collins 1979). Nevertheless, as we shall see, the belief in the link between innovation and economic growth, if specified correctly, also rests on some obvious truths.

We can begin with a wider explanation of the "great divergence"—in terms of economic history—to complement the narrower explanation of scientific and technological advance per se that was given in Chapter 1.[3] Some recent comparative-historical analyses have become skeptical about identifying a unique pattern that sets a European dynamic of economic growth (and thus technological progress) apart from other parts of the world before the nineteenth century (Hobson 2004; Pomeranz 2000). If this path is taken, however, then the unique role of science and technology in economic takeoff simply gets pushed forward to what Pomeranz himself calls "the great divergence" at the beginning of the nineteenth century, with a unique role of science and technology in the periods of the first Industrial Revolution in England/Britain which led to exceptional growth and paved the way for the second industrial revolution with more "systematic" and sustained economic growth.[4]

Mokyr has argued that this "sustained technological progress" resulted from "Western technological creativity," which, in turn, "rested on two foundations: a materialistic pragmatism based on the belief that the manipulation of nature in the service of economic welfare was acceptable, indeed commendable behaviour, and the continuous competition between political units for political and economic hegemony" (1990: 302). This explanation fits well with much of the thinking in current comparative historical sociology, but it should not be mistaken for Western "triumphalism." The technological/economic breakthrough was a one-off conjunction of factors in history; it does not "belong" to the West in any sense (cf. Goldstone 2002; Hall 2001). Once instituted, however, as Mokyr points out, "as long as *some* societies remain creative, others will eventually be dragged along" (1990: 302, emphasis in the original). And *this* pattern, ongoing technological progress throughout industrialized societies, is one that we are still inescapably subject to, not only in relation to economic growth, as for Mokyr, but for all the consequences described here.

Before we turn to these, however, it is worth stressing Mokyr's key point, that in the period from 1750 to 1914, "technological change . . . accounted for *sustained* growth. It was not caused by economic growth, it caused it" (1990: 148, emphasis in the original; see also 2002: 20, 29).[5] In view of the argument made in Chapter 2 that research technology is a precondition for scientific advance, an argument is needed here about the dissemination of this technology in society. This is readily provided by Inkster, who adds to Mokyr's ideas by emphasizing the role played by the strata who were the carriers of technological knowledge. Mokyr's and Inkster's explanations are complementary; the former points out that these carriers needed to have a certain degree of autonomy from the ruling groups amidst political competition in order to be able to break with tradition, while Inkster notes that they simultaneously needed to have the organizational means to transmit their knowledge. Thus Mokyr argues that no single political elite was able to block change (Mokyr 1990: esp.186–90), while Inkster describes the mobility of the stratum of carriers of scientific and technical information in terms of a "transfer of mental capital" in the period leading up to the industrial revolution (1991a: 32–59). Note that, in keeping with the argument here that "disenchantment" is the social side of "advance," this is a Weberian explanation inasmuch as it focuses on culture (or rather, in this case, knowledge) carriers, political blockages to culture, and the enabling preconditions for the rise of modern (rational) technology in the West. Moreover, this explanation leads into the account of the increasing autonomy of science and its strength as a profession that was given in Chapter 2 (though, as we saw, this autonomy was only institutionalized in the late nineteenth century).

Mokyr also makes an important distinction that we want to keep in mind in what follows, one that fits well with the distinction made in Chapter 3 between the period before and after the "research revolution": Mokyr distinguishes between the first Industrial Revolution, when major improvements were made, but only in some sectors of the economy, whereas during the second Industrial Revolution, science and technology were more systematically linked, and "the complexity of technological systems" increased (1990: 111, 113). In this context it is important to present Mokyr's definition of technological progress, which is "any change in the application of information to the production process in such a way as to increase efficiency, resulting either in the production of a given output with fewer resources (i.e., lower costs), or the production of better or new products" (1990: 6). This definition is useful, as long as "information" is taken to pertain to scientific and technological advance as they have been defined here (that is, relating to the physical world and to the environment, rather than including organizational changes, as they do for Mokyr). Mokyr's definition also points to the limitation of focusing on economic growth: while his definition allows us to explain economic growth, scientific and technological change also entails, as Mokyr is aware (for example,

2002: 163–215), changes in economic everyday life on the production and consumption side—apart from the impact on growth.

Aside from the unavoidable but not, in the end, problematic, "pulling up by one's own bootstraps" (or functionalist or circular) nature of the argument about the unique link between science, technology, and economic growth that we met earlier, it is essential, in other words, to add to growth the social changes apart from growth—the disenchanting consequences of scientific and technological advance. Put differently, we will need to specify the social changes that have taken place as a result of this advance—in the displacement of other institutions and other beliefs, and adding to the existing artifacts in everyday life.

Thus, again, we must have a clear comparative-historical and sociological account of the scientific (and technological) revolution, rather than a philosophical or narrative historical one. Modern science has a new legitimacy—it constitutes the modern world's dominant belief system—not because of ideas or culture that are separate from society, but because of what science, together with technology, does: a stream of research technologies successfully manipulate the world, lead to a more powerful intervention in the world, and migrate out of the laboratory and into our everyday lives. And on the basis of this impact, the idea of permanent change becomes deeply and widely held as the normal (or routine) pattern of social change in respect to science and technology. And although the argument here acknowledges the validity of this belief system, from a social scientific point of view, the term *advance* (in the value-neutral way in which it is used here) will be used instead of the value-laden but commonly used term *progress*.

The mechanisms whereby innovations translate into use and displace what preceded them have been much debated. An anthropologist provides a bird's-eye view: "Contingent elements, path dependency, and variants of adoptive behaviour all converge virtually to refute the possibility of order and regularity in technology diffusion. Yet it is evident that superior technologies . . . have always spread irresistibly. Disturbances and impediments of various kinds may deflect or delay, but do not permanently block, technological advance" (Adams 1996: 29). Again, when we examine the process of diffusion more concretely in Chapter 6 on consumption, we will see that, although the process is uneven, the important point is that the consequences of technological advance are homogeneous, even if they simultaneously create greater diversification.

We can consider another interesting perspective, this time from a historian, who points out that,

> there is, in general, no social prestige in employing a technology that has reached a stage of maturity or decline. However, there is often high social status in employing a technology that is in its initial stage of growth or—paradoxically—a technology which has already declined and been by and large abandoned. It is, for example, prestigious to write either on a portable word-processor or with a fountain pen. A ball-point pen, *le dernier cri* in the 50s, lends little glamour to its present owner. (Lindqvist 1994: 284)

The example is instructive since it highlights—again—that many technologies that are in routine use have become invisible. Notice also that this is a consumer item, where choice, within limits, is possible, and yet certain tools become standard. In production, and in many areas of consumption, choice is therefore irrelevant, since advance is, if not obligatory, then nevertheless the norm.

The historical perspective fits well with the argument made here. Lindqvist argues that we put too much emphasis on S-curves of adoption[6] and so exaggerate growth without also taking the decline of technological systems into account: "The variable in the development of technologies that ought to concern us historians"—and, I would add, sociologists—"is the prevailing *technological volume*, i.e., the sheer amount of existing technologies at a given time" (Lindqvist 1994: 277). I would add that the only way to make sense of this volume is through the notion of a "footprint" as I have used it here, including the footprint of cumulative consumption that will be explored later.

From the American System to Mass Consumption

With these ideas about the diffusion of technology in mind, we can return to the relationship between modern science, technology, and economic growth. The link between new technologies for harnessing energy during the Industrial Revolution has been mentioned in the context of infrastructures in Chapter 3. Two subsequent shifts stand out: the advent of mass production, stretching out according to Hounshell over the period 1800–1932, but concentrated in the second industrial revolution of the late nineteenth century and the period leading up to World War I. The second (which we will come to in a moment) is the postwar diffusion of American large-scale R&D and business practices.

The shift toward a coupling of R&D to production methods during the second industrial revolution, as we have just seen, put production on the course of more *systematic* growth in Mokyr's view. This point is reinforced by Hounshell, who notes that the advent of mass production was simultaneously a technological and economic transformation: "The historian might quibble with definitions of mass production—whether it is a doctrine, a business philosophy, a large production output, or a technological system . . . mass production is all of these put together" (Hounshell 1984: 122). Howsoever it is conceived, he continues, this was a qualitative leap in the way in which, with the use of electrically powered machinery, goods were now produced by more effective means and on a different scale: Thus "the bicycle and automobile industries, with the help of the machine tool industry, would complete what the firearms and sewing machine makers had begun" (Hounshell 1984: 122).

Hounshell depicts a broad change in the world of production, but even in the historical detail of the processes that he describes, we can see how the Weberian definition of technology used here, as an increasing rationalization of the methods of production, applies to each of the areas that were part of the emergence of mass production. We can also anticipate by leaping ahead to the present day: Rationalization of the type described by Hounshell in mass production may no longer be so central since factory work has become a minority occupation. The era of mass manufacture revolved around the production of existing goods on an unprecedented scale, whereas today, the economy revolves more around creating growth through products that did not previously exist—something that will shortly bring us to Schumpeter's extension of Weber's ideas.

But, as with the argument that I made earlier about the Industrial Revolution, the separate impact of science and technology must be stressed: again, to rely on Hounshell, this was a harnessing of machines to power.[7] And, as we saw earlier, big science and large technological systems were themselves products of this new form of economic organization: the *tightening* of the relation between science, technology, and the economy that characterizes the period after the advent of the "American system" can be defined as just this increasing mutual dependence between them. Mokyr sees this tightening relationship as characteristic of modernity: "One way of describing the modern age is that the relative importance of knowledge for its own sake has declined relative to knowledge that may be mapped into better techniques" (2002: 287).

A good example of this tightening, or the simultaneous emergence of new technologies and the increase in scale and scope of the economy, to complement

Hounshell's account of mass production, is Beniger's (1986) explanation of the evolution of "technologies of control" to cope with emerging mass consumption, an array of feedback loops whereby this demand could be met with new logistical tools for the distribution of goods as well as tools for marketing.[8] Another is Yates's (1989) account of new office technologies for the storage, retrieval, and dissemination of information, another necessary accompaniment allowing the processes of mass production to be managed effectively on a vastly extended scale.[9] Note that purely economic—"pull"—explanations, whereby economic changes necessitate the development of new scientific knowledge and technological artifacts, will always be is insufficient here. What is required is that scientific and technological advances open up new possibilities and impose new constraints to cope with larger markets.

Innovation and Markets, Networks and Monopolization

Yet markets, again, within the analytical framework proposed here, do not exist in the abstract, but need to be tied to the diffusion of the new products of science and technology. As Collins puts it, "socially and economically, an innovation cannot really be said to exist until it is used" (1986: 80). This diffusion undergoes a qualitative shift with the American system because innovation itself, as Collins argues, becomes organized in a factory- or mass-production-like way: "The activities even of academic laboratories—which themselves resemble little factories, repeating experiments endlessly, with careful measurements and systematic variations—may well owe more to organizational technique than to [scientific (my addition)] theory per se. In fact, they are laboratory *factories*. In this light, it should not surprise . . . that technology may have always had more influence over science than the other way around" (Collins 1986: 113–4). Collins's approach to scientific and technological change allows us make the coupling back from "use" to "scientific and technological advance" (even though Collins includes organizational change in this, which I do not). But excluding organizational advance allows us to avoid, on the one hand, abstractly introducing scientific and technological advance as exogenous factors of production, and on the other, not to *include* treating organizational changes that are produced by science and technology unless they are the result of the concrete diffusion of scientificity into organizations and adopting new artifacts as tools to cope with organizations' changing needs in relation to their environments (together, the rationalization-disenchantment and exoskeleton-footprint of production). Hence, once again, we can close the loop by concluding, with Collins, that "rapid social diffusion"—of new technologies, in my view—"is exactly what we mean by a business boom" (1986: 84); in other words, that periods of exceptional growth occur when new scientific methods and technologies diffuse more quickly.

A similar argument is made by Chandler, but markets in this case are put into a historical context: "It was the development of new technologies and the opening of new markets, which resulted in economies of scale and scope and in reduced transaction costs, that made the large multiunit industrial enterprise come when it did, where it did, and the way it did" (Chandler 1990: 18). It should be noted here that "new markets" mainly meant conquering America's geographical reach; in other words, what I have called here extending the environmental footprint of large technological systems. Apart from the functionalist logic of Chandler's argument, which we have already encountered, for him, too (as for Hounshell, Beniger, and Yates), the focus in the end is not on geography but on technology: "It was not until the 1870s, with the completion of the modern transportation and communication networks—the railroad, telegraph, steamship and cable—and of the organizational and technological innovations to operate them as integrated systems, that materials could flow into a factory or processing plant and finished goods move out at a rate of speed and volume and with the precise timing required to achieve substantial economies of throughput" (Chandler 1990: 26).

The limits of the extent of the market therefore lie at the nexus between mass production and the mass-consumer economy, and thus at the extent to which economic growth feeds back into innovation, a limit that is at once open and subject to constraints. Lundgren has argued along the same lines that technological innovation and industrial networks evolve together in a constant feedback loop: "The industrial network is driven by concurrent sociotechnical processes of creation, integration and expansion, where as the evolution of the technical system is promoted by the creation of novelty, development, and use" (1995: 190). This is the mechanism that governs an economy in which new technology is used to create new forms of demand, and it is also where Schumpeter comes in (Schumpeter 1934, [1942] 1994), especially if we apply his ideas to how entrepreneurs use technological or scientific means to create new combinations and thus stimulate new demands. A Schumpeterian analysis entails a shift from looking at economic growth and rationalization on the macrolevel—to the more mesolevel of social mechanisms by which markets work. And here, Weberian and Schumpeterian analyses have lost none of their force: "markets" work by the "always fleeting" monopolization of opportunities. They work, to put it in the language of more recent economic sociology, through the exploitation of new niches.

Schumpeter thinks that new niches or closed opportunities are exploited by entrepreneurs who create new combinations. We can add to this the structures within which they do so, the firms which, according to Harrison White (1981), constantly monitor each other to see where advantages can best be gained (or such new niches can be found). As a final element, we need to add the changing context, from the perspective of the shifts sketched here, of the environments in which the firms do this monitoring. This context is the transformation from one dominant scientific and technological (again, the two are increasingly linked) area of advance to the next, and the mechanism by which this transformation occurs, which in turn, according to Schumpeter (following Collins's exposition; 1986), is a flow into and concentration of financial resources in areas where profits are expected to be greatest. The new combinations of Schumpeter's entrepreneurs—if they are based on scientific or technological innovations—therefore take the form of identifying and seizing the niches in which innovations and financial resources combine to bring advantage. These new combinations can also be regarded as monopolizations of opportunities—or lock-ins (Arthur 1989) and other means of temporarily "locking out" the competition in a particular niche.

Thus, from a Weberian and Schumpeterian perspective, the networks that Lundgren (1995) describes are not only alliances between different groups that bring their combined resources to bear on innovation, but temporary Schumpeterian monopolizations—or stabilizations in markets, as Fligstein has put it more recently (2001: esp. 90–91)—which exclude competitors. Or, to put it in Weberian rather than network terms, innovations are a form of market closure, or an appropriation of opportunities (technological and scientific, and within a new market niche) by powerful groups that are able to exclude others. Note how this account is consistent with the notion used here of scientific and technological *advance*—novelty needs to be (and can only be) temporarily monopolized to bring advantage in networks. We should notice here, too, as in Chapter 3, that neither a conflict nor a functionalist perspective can do full justice to new technologies in markets: the conflict perspective allows us to identify "exclusion," but exclusion in this case is temporary and non-zero-sum. Functionalism allows for the open-ended expansion of markets to meet needs but imposes no constraints on this expansion. Networks, which consist of exclusion but reach only as far as novel (and open-ended) "use," thus constitute the most sophisticated analysis of innovation that is currently available (beyond conflict and functionalism), since the feedback loop from use back to the creation of novelty has not yet been systematically analyzed. And again, we need to remember the other side of advance: what has just been said applies to innovation from the production side, how firms innovate to create new products. Cawson, Haddon, and Miles (1995) look at the other side, how firms feed user needs into their products through research on consumption. But while there is an extensive academic literature on "innovation" in economics, how a stream of new technologies comes from R&D labs into the home and other places is not well researched (though the process has been sketched in Chapter 2).

The Diffusion of the American System, National Harnessing, and the Limits to Growth

With this, we can turn to the second major shift. After World War II, and after a period in which the economy had been less international and concentrated on the war effort, came a further tightening—or perhaps we should say "systematic harnessing" in this case: In the postwar period, as Whitley says, "science became seen more as a factor of production than just one component of 'high culture'" (2000: 296). And in van der Wee's view, "the systematic and large-scale organization of industrial research after the Second World War determined to a high degree the productivity of labour and capital in the West" (1987: 202). He points out that "the most successful of these science-related, research-intensive industries after the war were space exploration, electronics, pharmaceuticals, chemicals, petrochemicals, and precision instruments. It was precisely these high-technology sectors that achieved the highest growth rates. Moreover, by means of their technological predominance, they 'colonized' the more traditional industries" (1987: 205). He also describes another form of colonization that fits with the ideas described earlier: the transplantation of the American system to Europe and beyond. Van der Wee argues that "it is difficult to overestimate the role played by American multinational corporations in reducing the technological gap between industrialized countries during the 1950s and especially the 1960s" (1987: 213).[10]

Harnessing science and technology to the *national* economy was at that time a new way of deliberately coupling R&D to the economy, and this process is still ongoing today, as researchers, who need to demonstrate their "usefulness" to state funding bodies, are well aware. Thus Whitley notes that there is an "increasing dependence on external funding of programmes and projects in most scientific fields" (2000: 290; see also 276, 289–90).

In this context of the macrolevel relationship between science and technology and the national economy, we must return again to the concept of national systems of innovation. As Whitley has noted, the national systems of innovation literature has focused more on technical change than on support for scientific research (2000: xxii). He also points out that the difference between these national systems continues to be important, contrasting, for example, the decentralized and more demand-led system in the United States with the more bureaucratically steered systems in Europe and Japan (2000: xxii–xxxi). But again, from the perspective advanced here, the details of policy can be left to one side; the important point is that all national systems are geared toward capitalizing on the advance of science and technology. Especially in view of our Swedish-American comparison, or from surveying the developed nation-states as a whole, we should bear in mind that national systems of innovation share many of the same basic features (Drori et al. 2003: esp. 155–73). Moreover, much of the literature in this area is directed at policy makers who should steer the process of innovation, but one question that was mentioned earlier in passing can be asked again in this context: namely, to what extent can innovation be steered at all on a national level?

Debates about innovation and economic growth take place in a variety of disciplines and subdisciplines and bear on important practical issues. They are sometimes about innovation and economic growth at the national level and at other times narrowly focused on particular sectors or institutions for particular times and places. One shortcoming in these debates is that we also need a synthetic perspective on innovation and economic change, one in which scientific and technological advance do not become submerged among other phenomena (including "growth") *and* which identifies the long-term dynamics and relations between science and technology and other parts of the social world.[11]

On this larger level, the national harnessing of research has been obscured by the alleged multinationalism of the American corporation, but as Pavitt and Patel (1999) point out, there are persistent national patterns of organizing research (or patterns that several clusters of nations fall into). So we have two simultaneous postwar developments: the spread or diffusion of American R&D and corporate organization as a norm—and "national" systems. But it is a mistake, as Weiss has argued (2003), to see internationalization and the strengthening of national systems as zero-sum; in this case, they occur side by side. And finally, these two postwar developments apply to all four types of scientific/technological institutions—university, government/military, industrial, and public research organizations. Yet apart from the changes in the relationships between these four, the main gear shift, again, is that all four of them have become more closely coupled to economic need: "Scientific research in general is increasingly an activity to be invested in, directed and organized on a national and international level" (Whitley 2000: 289). Again, today we take this close coupling for granted, but we should remember that this was not so in the pre–World War I period, when there was only a loose coupling—or science and technology were more autonomous and still more tied to military-administrative rather than economic needs.

We also need to put the postwar period in a longer-term perspective in another way. There is widespread agreement that the *rate* of contribution of R&D to economic growth decreased in the last third of the twentieth century—after a period of world-historically high rates during the Golden Age (1950s to 1970s). According to Maddison, "technical progress has slowed down. It was a good deal faster from 1913 to 1973 than it has been since" (Maddison 2001: 25).[12] This may be why the Durkheimian truth (or the legitimacy) of the belief that science and technology lead to economic growth may now be fraying around the edges, just as resources devoted to R&D are perhaps plateauing.

America and Sweden: Leading Edge, Catching Up, and Convergence

The focus so far has been on science and technology in American society because America was at the leading edge of the transformation of science and technology in the twentieth century. But it is important to ask about the importance that can be ascribed to the American pattern; for example, whether other industrialized societies have converged or departed from the American pattern. This "exemplariness" immediately needs to be put into context: Gellner has noted, for example, that the "greatest elective affinity" of the scientific world view "need not be, and probably isn't, with its place of origin. The first industrial and scientific nation [England/Britain] is not, at present, at the top of the First Industrial Division" (1992: 61). Inkster makes a similar point about the United States emerging as the main economic and technological superpower after World War II, but soon being overtaken by the powers that were defeated in the war in terms of innovation (Inkster 1991b).[13] Scientific and technological leadership is fleeting, especially since, as argued earlier, science and technology are universal and not socially shaped in the way that this has been argued in the sociology of science and technology. Put differently, science and technology are oblivious to who or what its carriers are. Further, even if America is still at the leading edge in certain areas, it is difficult to argue that it contributes "more" to scientific and technological advance than do other developed societies.

This is a good place to note that science and technology per se do not cause differences in (distributive) power. The preeminence of America as a world power does not rest on the characteristics of science and technology alone (Hounshell 1996: 41). Moreover, as argued earlier, American military advantages stemming from science and technology are, in principle, available to all. This leads to a more general point about science/technology and social power, which is that in the area of technological progress, we are not just concerned with the distributive power but also with collective power—that is, not the power of A over B, but power "whereby persons in cooperation can enhance their joint power over third parties or over nature" (Mann 1986: 6, drawing on Talcott Parsons's ideas about power).[14]

How then are science and technology integral to modernization? To examine this, we can briefly look at Sweden in relation to America. Sweden is a good example because it satisfies the criteria of the variation-finding comparative method (Tilly 1984: 116–24), that is, in other respects, it lies at the other end of the range of "advanced societies": in its political system, economic system, economic lateness or "backwardness,"[15] and presumably in its culture and mix of institutions (education system, factor endowments) related to science and technology. It also has a well-researched history of science and technology.

Sweden and America have nevertheless undergone similar changes in the makeup of the workforce with the shift from agriculture to services (compare Hult, Lindqvist, Odelberg and Rydberg 1989: 329 and Beniger 1986: 21–26). Sweden does, of course, have a unique technological profile: machine tools have played a larger role in Swedish industry than anywhere else (Hult et al.

1989: 244, 320), which in turn has even deeper roots in the prominent role of ironworks in Swedish history. But again, a key question is: how much importance should we attach to such distinctive historical patterns from the point of view of today's diversified or "mixed" economies?[16]

How then should we compare the two countries? First, there has been catch-up. This seems an obvious point, but it is not well theorized: there are comparative studies of science, technology, and economic growth and of political economy,[17] but these overlook long-term convergence. A more well-rounded picture would include the idea that modernization is a functional requirement of industrialized societies—that technological infrastructures, large-scale R&D, mass production and consumption, and not just "innovation," are the needs of developed societies (even if this need not commit us to a broader convergence outside of the harnessing of science and technology to the economy).

The American pattern, as we have seen, had a wider significance because, after World War II, it was seen as a "reference society" by others countries. This is partly "ideology" (or culture) since science and technology, as argued earlier, are universal. But from a longer-term perspective America has been regarded as a model just like other leading-edge powers were regarded as models during other periods. One example, as Rosenberg (2004) points out, is that even today, the role of the research university in American society is seen as a model throughout the developed world (which Sweden has tried to emulate in fostering academic enterpreneurialism; see Elzinga 1993). This mix of catch-up, functional modernization, and emulation is best pursued, however, in an area where there is often thought to be a great divergence, in culture and consumption (in Chapter 6). First, we can examine the role of technology in the second major sphere apart from economics, the sphere of politics.

5   The Mediation of Politics as a
Large Technological System

THE CONCEPT OF LARGE TECHNOLOGICAL SYSTEMS HAS MAINLY BEEN applied to society's infrastructures: energy, transportation, and communication. In this chapter, this concept will be applied to the role of information and communication technologies (ICTs) in political change. This will provide a further concrete example of the implications of large technological systems (discussed in Chapter 3) and at the same time extend the overall argument about technological determinism to the sphere of politics. But there are several additional reasons for examining the role of technology in politics: one is that an argument will be made that the *range* of ICTs should be considered as a whole—as a system—in order to gauge their political impact. And the second is that the specific role of technology—its consequences—are mostly ignored in discussions of the role of media in politics. And the third and final reason is that new media technologies, and especially the Internet, have arguably changed the role of ICTs in politics, and this claim deserves close attention in light of the claims of this book about technology and social change.

Again, the analysis is based on two countries that have political and media systems that are as different as they can be in the political sociology of developed societies, Sweden and the United States, but here the focus is on the role of ICTs in shaping the public sphere. The emphasis will be on print, radio, and television in political life—though the Internet will be added even if it is still too early to assess its impact. The question of the role of technology in social change is rarely discussed explicitly in media studies and political communication (though it is often implicit). More specifically, we will examine an implication that follows from the technological determinist position put forward here: whether the effect of technologies for politics in different countries contributes to convergence between them.

The conclusion can be anticipated: that the implication of convergence follows, albeit with some qualifications. In other words, technology has played a rather similar role in these two politically quite different countries, and their mediated public spheres have therefore also converged—becoming more diversified but also congealing into a set of dominant technology uses and forms of mediated politics in both countries. The qualification is that in other respects the two political cultures remain as distinct as ever, from which it follows that ICTs do not have a single global or homogenizing effect on politics that eliminates these differences.

Before plunging into this topic, it is useful to expand briefly on why it is necessary to revisit the well-trodden topic of the role of the media in politics.[1] One is that discussions of media and politics typically only deal with a single medium, such as television, or they examine a single country or a limited time period. Few treatments address the whole range of ICTs or examine the comparative-historical evidence over the long run. Further, political scientists and media scholars rarely address the role of *technology* in political change specifically. And even if they include the role of new technologies in their analyses, they often lack the conceptual tools for analyzing why different technologies have different effects—take, for example, the difference between broadcast television as against cable and satellite television with its fragmentation and proliferation of audiences. Moreover, media scholars and political scientists often generalize about the effects of new technologies without relating them to everyday uses.

Sociologists of technology, on the other hand, have developed some useful concepts in the study of the social implications of new technologies, but they have not dealt specifically with media or politics (exceptions are discussed later). Nor has the concept of large technological systems, developed in the sociology of technology, been applied to understanding the political role of ICTs. Moreover, this chapter, like Chapter 6, will need to address the criticism that studies of effects do not actually take into account the everyday uses of technologies.[2] To address this shortcoming, the uses of ICTs will be tracked here in terms of the historical phases in which they dominated and how they have been used in mediating politics.[3] Hence a number of elements will need to be integrated—including studies that contain material about the uses of technologies in the two countries over a long period, comparative media studies, political sociology, and the sociology of technology. This combination will yield a comprehensive account of the role of technology in relation to media and politics.

A final reason to revisit the topic of politics and the media is to put the recent debate about the role of new media—and the Internet in particular—into a broader perspective. So it has been argued that new media, for example, with their possibilities for greater interactivity, more direct access, and the formation of new communities, open up new political opportunities. Opinions remain divided, however, as to whether the Internet pushes the mediation of politics in new directions, or if this new tool merely fulfills the same functions as existing ones. Both positions, as we shall see, have limitations.

This chapter is organized as follows: the first section outlines different perspectives on the media and politics and provides the rationale for the approach adopted here, which is to focus on how media systems as large technological systems mediate between political and media elites on the one hand and civil society on the other. The next section reviews a number of studies about the role of ICTs in Sweden and America, and describes how technology has contributed to convergence. More specifically, two studies will be assessed that compare the two countries in the light of the argument developed here. Some brief comments about the Internet follow, and the conclusion draws out the implications for "mediated politics."

Politics and the Media

At the center of debate about ICTs and politics has been the notion of the "public sphere." A number of key questions have been framed around this concept: Has there been a fragmentation or homogenization of the public sphere? Second, do new media technologies lead to more—or less—input of the public into the political process? Put differently, do they enable more—or reduce—responsiveness from those who govern? All three questions revolve around whether changes in the public sphere have strengthened or weakened democracy. A fourth and less common question is whether new technologies lead to greater convergence between different political and media systems, and if so, what the implications of this convergence are? This chapter will concentrate on the last question, but this will bear importantly on the answers to the first three.

On the side of social theory about the media and politics (again, where technology is largely absent), three main perspectives are available: critical theory, which counterposes an emancipatory public sphere against growing political and economic power that encroach upon ("colonize") and diminish it; in other words, the media are becoming less autonomous. This position argues that the democratic role that an autonomous public sphere once had in encouraging political input, in Enlightenment Europe, has declined and should be revived (Habermas 1982: esp. vol. II 420–44). A second, functionalist perspective, posits the opposite—an increasing differentiation or autonomy of a separate sphere of communication within society by means of which values can be integrated as an input into the political system. In this case America, which has the greatest separation between the media system and the political system, is often regarded as a desirable model (see Hallin and Mancini 2004: 78–9; as we shall see, Luhmann 2000 provides an important variant of functionalism). A third, postmodern or constructivist perspective, argues that the media increasingly create a reality of their own with its own separate logic, with the effect on politics that this "construction" of political reality by the medium is inescapable.[4] Postmodernists and constructivists go on to argue that this construction is inevitably dependent on a particular cultural context and, in the end, once this—rather arbitrary—cultural context is laid bare, it can be opened up to contestation.[5]

A fourth—conflict—perspective could be added here, though it is a theoretical perspective within political sociology and has not (to my knowledge) been extended to the role of media in politics. In this case, instead of the public sphere of the media, the focus is rather on the contest between elites and civil society, including social movements (Collins 1999: 35–38; Mann 1999: 254–57). The role of the media in politics could be built into this picture by adding media elites to political elites. The power of the media in this case is balanced between these two elites on the one hand and a broader (civil society) public on the other, with the media playing an important role in a larger ongoing struggle to further democratize society "from below." To anticipate my conclusion about these perspectives, I will argue that the truth lies with a balance between the conflict and consensus (functionalist) roles of the media and that this balance differs between the two countries considered here, but further that the large technological media systems common to both countries are locked into a momentum of their own—with a mixed effect on their democratic potential.

Irrespective of these theoretical perspectives, the aim of analyzing the role of the media in politics should be to gauge to what the extent the media play a role in the mutual shaping between those who wield political (or state) power and their citizen-publics.[6] Or, as Starr describes this relationship for the time when mass media first began to play a role for society as a whole, "rulers . . . acquire[d] new means of monitoring society . . . [and] the public . . . began to acquire new means of monitoring their rulers. Society [is made] . . . more legible to the state . . . [and] the state . . . more legible to the public" (2004: 45). With this, we can leave the extensive debates about the public sphere and civil society to one side for the moment.

Information and Communication Technologies as a Large Technological—Media—System

The range of media or ICTs and their cumulative or aggregate effect can be regarded as a large technological system. As we have seen, the concept of large technological systems is associated with Hughes, who argued that infrastructures of technologies, for example the electricity grid, combining technological and social elements, become congealed in particular ways and then exercise their shaping of society *as a system* (1987, see also Summerton 1994).[7] The advantage of treating several ICTs as a system in this case is not only that this includes the technological and social sides of the system, but also that it allows us to treat this range of media as a whole even though the system clearly consists of a combination of different media channels or types of media.[8]

What then characterizes *media* systems? On the technological side, they consist of the range of ICTs—print, radio, television, and the Internet.[9] On the social side, they include government regulation, the political economy of ICTs, and their distribution networks. The key feature of large technological systems is that the two sides intertwine (though this is not a reason for not keeping them analytically distinct). Put differently, the large technological media system consists of how the technology for the mediation of politics—or for political communication and information—intertwines with the political and economic framework that governs this technology.

The various parts of the media system took time to add up to their present-day shape. Newspapers and telegraphy already had a major impact in the seventeenth and eighteenth centuries, but inasmuch as this chapter focuses on mass citizen-publics, it makes sense to start with the press in the late nineteenth century since this was "the first medium capable of reaching a mass audience" (Ward 1989: 21). This was also the period in which the technological systems—the infrastructures—for large-scale printing and news distribution (via telegraphy) were extended in scale and scope. To this were added radio broadcasting in the 1920s and television broadcasting in the 1950s. Again, it is important to regard these three technologies (and now also nonbroadcast digital radio and television and the Internet) as part of one large technological media system because the patterns that we are interested in—how the system mediates between elites and the public—are features of the system as a whole.

One advantage of applying the concept of a large technological system to media is that the focus is not on individual attitudes to particular media (as with survey-based research; see, for example, Norris 2000), but rather on how the system as a whole enables or constrains those who control and are affected by it. This also allows us to recognize that the three groups identified earlier have asymmetrical relations to the media system: political elites depend on their citizen-publics for support via the media system (but not vice versa), and both political elites and citizen-publics have increasingly come to depend on media elites. And although media elites see themselves and the media merely as a conduit or mirror of public opinion, they in fact control access to this system (see Gans [1979] 2004, and below). And finally, as we shall see, the media system has gained increasing autonomy at the same time that political and media elites—and citizen-publics, and especially the social movements in civil society—have sought to harness the media system more strategically.

Another benefit of regarding media systems as large technological systems is that this allows us to put new technologies into perspective: a new technology like the Internet may introduce a new dynamic into the system, but it must also "fit into" an existing congealed system. For example, on the technological side, the Internet has been grafted onto the telephone network and subsequently also cable and wireless networks, and on the social side it has had to be made compatible, for example, with existing patterns of telecommunications policy and market regulation. This fitting in, as we shall see, applied just as much to radio and to television.

A final advantage is that the *content* or output of the system, the combined technological capabilities and constraints for conveying content and how the attention space is distributed across different media, can also be treated as a whole. This will be important because one argument that will be made is that with the cumulative impact of more media, there is nowadays a greater diversity of channels and of content, but this content is also increasingly managed at the "sender" end, and there is also a limited space among "senders" and limited attention space at the receiver end. This is another structure into which new media technologies, such as very recent trends toward user-distributed and produced ("user-generated") content via the Internet, need to fit.

Two Political Systems—Two Media Systems

Politically, Sweden and the United States occupy two extreme poles in the comparative political sociology of advanced societies—on the left and right of the spectrum, or the social democratic and "Anglo-American" liberal ends if this terminology is preferred.[10] This shapes the two media systems, though as we shall see, it is more complicated than to contrast an interventionist state with a laissez-faire liberal one.

Swedish politics in the twentieth century has been dominated by the Social Democratic party. Yet, as Åsard and Bennett point out, in all this time they have only won two elections with an outright majority of parliamentary seats, and so have constantly been forced to adopt a course of compromise rather than confrontation with other parties (1997: 88). This "pragmatic view of the role of the state" (1997: 90), according to Åsard and Bennett, reaches back deeply into Sweden's liberal nineteenth-century past.

The culture of the Swedish political system in the twentieth century is encapsulated in the notion of the "folkhem," or people's home. Åsard and Bennett describe this as consisting of "three basic ideological components . . . active government . . . a general welfare policy . . . [and] corporatism" (1997: 98). The strong role of the state, with its responsibility for the welfare of its citizens and for maintaining a cohesive and well-functioning society, has continued to be a deeply entrenched and widely shared ideal in Sweden. This again puts Sweden at the other end of the spectrum in comparative sociology to the United States with its tradition of ensuring individual freedom *against* the encroachment of the state.

Despite the strong role of the state, the Swedish media system is in fact characterized by a mixture of liberal independence *from*—as well as an active role *by*—the state. Sweden has one of the longest-standing traditions of guaranteed press freedoms in the world: The Freedom of the Press Act became part of the constitution in 1766, stimulating political debate in the newspapers and making the parliamentary (Riksdag) proceedings public under the stipulation of freedom of information. At the same time, the state has, from the beginning, played an active role in shaping the media, especially from the late nineteenth century onwards when it had to ensure that diversity was maintained among newspapers in the face of mass party politics.

The first mass medium, the newspapers, were strongly tied to Swedish political parties, including their financial backing—as elsewhere in continental Europe. The connection between political parties and journalists in the Swedish media was also strong by tradition, but the press has increasingly dealigned itself from political parties (Hadenius and Weibull 2003: 89). This relationship between them has been progressively weakened such that by the 1980s party financing for newspapers was marginal. Moreover, the party-political connection of journalists had been replaced by the notion of an impartial news journalism, partly "imported" from the United States (Hadenius and Weibull 2003: 290–98). This impartiality also has a long-standing base in journalists' professional organization, though journalists' self-image has undergone several transformations in the course of the twentieth century, with the 1960s and 1970s a particularly "pro-active" period (Hadenius and Weibull 2003: 318–20; see also Djerf-Pierre and Weibull 2001).

With radio broadcasting, the system came to be state owned, continuing the tradition of the postal and telegraph services. This tradition of the state being responsible for the infrastructure (that was described in Chapter 3) contrasts strongly with America, where the infrastructure was developed commercially—then carried over into television. The broadcasting system in Sweden, which developed only somewhat later than in the United States, was thus initially fostered by the state in the public service tradition. But the state's role was not so much, as in Britain, to provide a vehicle for public information and education, but rather to provide a technical network to cover the whole of the country. Moreover, the state's role has always been at one remove, providing a framework for mass media institutions and fostering universal access and regional coverage, but leaving operational autonomy and responsibility for content in the hands of the institutions (nowadays, Swedish television has two public stations and several radio stations) as long as they fulfilled their obligations to the public. Hence Hadenius and Weibull characterize the Swedish system as "regulated independence" (2003: 214–15); in other words, the state maintains its distance from the internal workings and political content of the system, but regulates them indirectly.

Since the 1980s, there has been a drift toward a more market-oriented rather than a public service-oriented system for broadcasting (Hadenius and Weibull 2003: 443). With the development of a mixed system of partly public service and partly commercial television and radio, there has been a debate that can be found elsewhere in Europe about whether public broadcast services should compete for large audiences with the commercial sector or whether they should offer the kind of more diverse programming for smaller audiences and for social and cultural purposes that commercial programming cannot reach (Hadenius and Weibull 2003: 250–53). In this debate, again as in the rest of Europe, no consensus or steady state has been reached (in America, public funding for broadcasting has played a comparatively marginal role).

Audiences for press, radio, and television, in terms of exposure and their preferences, have been remarkably stable in Sweden since the 1970s (Hadenius and Weibull 2003: 393–441). The main change is that commercial radio and television have taken a sizable share of the audience away from public service broadcasting, though public service content is still more trusted than that of commercial broadcasters (Hadenius and Weibull 2003: 438). Like elsewhere, however, television has "become ever more important as a means of orienting oneself in events in Sweden and the world" (Hadenius and Weibull 2003: 441). There has also been a shift away from a "common frame of reference" (Hadenius and Weibull 2003: 460), which existed when everyone had access to the same (public service) channels. This has weakened with the proliferation of channels and a proliferation of media in the home in recent decades (Hadenius and Weibull 2003: 256). The increasingly individualized listening and watching patterns of Swedes since the 1960s are evident in social histories of domestic media consumption: as in the United States, now that households for the most part have two or more television sets, radio, and also the Internet, household members can consume their news and other content separately and via several channels (Höijer 1998; see also Chapter 6).

The Swedish large technological—media—system therefore has several distinctive features: one is that, like its other large technological systems (the main ones being energy, transport, and communication), the state has taken the role of creating the technological infrastructure of the media system (Kajser 1994, 1999; and Chapter 3). Once in place, however, its role has been at arms length, providing a public service broadcast system that ensures universal access and regional coverage, and recently also allowing ever more commercial broadcasting so that the system is a mixed public service and market system. The balance has shifted more toward the market in recent decades, as it has in other systems that were public service oriented, but the role of the state in promoting a system that is balanced, accessible, and promotes the public good is generally unquestioned in Swedish society.

In the United States, as in Sweden, newspapers were, from the outset, regarded as the preserve of free expression and of critical political debate, though the emphasis from the start has been on safeguarding these as a counterweight to the state rather than on providing "transparency" as in the Swedish case. The newspapers were part of a burgeoning public sphere from the time of the American Revolution onwards, but reached a mass audience only with the penny papers in the second half of the nineteenth century. There was not only an intense competition for circulation in large cities, but also a market for newspapers on a more local level. Unlike in Sweden, however, the American newspaper system was more strongly shaped by the need to sustain itself commercially through advertising.

The second distinctive feature of the American media system before the era of broadcasting was that free expression, enshrined in the Constitution's First Amendment, was safeguarded by law and *against* the encroachment of the state.[11] The same applies to the role of law in preventing a monopoly of ownership; legal regulation of the market rather than state intervention was used to prevent monopolies in distribution channels and in geographical concentration—the state's role was not to direct or supplement the system or promote a public service outlet of its own. These legal guarantees were also thought to prevent economic power from unduly interfering in politics.

A key indicator for U.S. media has therefore been whether people place trust in the role of the media in safeguarding the freedom of the media. This issue intensified with the advent of competition from broadcasting. In the late 1930s, when the radio broadcast system had been in place for over a decade, radio was the most widely used source of news for most Americans and was, according to a contemporary survey, regarded as more accurate than newspapers (Czitrom 1982: 86). The role of most trusted medium shifted quickly to television after this new medium had, in turn, displaced radio as the dominant and most trusted source of the news.[12] And although newspapers are still an important medium, sales have been falling in the United States in recent decades, and television is a "far more commercially dominated" medium (Norris 2000: 281).

Before comparing the two countries in more detail, we can briefly identify the turning point between the time before a national media system was in place—and after—by reference to two in-depth sociological studies of comparable towns in the United States and Sweden that were undertaken in the 1920s and 1930s, *Middletown* and *Medelby* (this comparison will be deepened in Chapter 6). What both studies demonstrate is that a shift had taken place in the late nineteenth century: in America's Middletown, a "torrent of printed matter" (Lynd and Lynd 1929: 232) had entered people's lives at the end of the nineteenth century where there had previously been mainly religious reading matter. Another shift was that out-of-town newspapers—state and national—had been added where before there had been almost exclusively local newspapers (1929: 472).

The same applies to the Swedish Medelby. In this community, over half of the households read a newspaper by the late 1930s (Allwood and Ranemark 1943: 241), and regional and national newspapers had become an important addition to local news, especially—as in Middletown—among a more elite readership. The authors of *Medelby* also point out that the printed word still carried more weight than radio news (1943: 243), unlike in the United States. But the main finding of both studies that is relevant for mediated politics is that, although the authors of both studies were concerned about the political processes for reasons *apart* from those having to do with the media, they agree that a new window on the world had been opened for both towns, and a national perspective on current events had been created where previously the focus had been local. Print—and then radio—had become consolidated into a national system.[13]

The addition of television in the period after these two studies then intensified this shift, but what changed with radio and television was also that the significance of print was increasingly overshadowed, especially by television. The United States has experienced a more extreme version of the Swedish proliferation of media outlets: Up to the 1980s, television was dominated by three national networks, but by the end of the 1990s almost every household with a television had access to a plethora of cable channels (Bimber 2003: 83–86). So that whereas the great majority—Norris says 90 percent—of the American television audience watched the news on the three networks in the 1960s, after the proliferation of channels less than half do so now (or did in 1999) (2000:

100). Norris suggests that the "fragmentation of [TV] channels . . . has gone further in the United States than in most OECD countries, although others have been quickly catching up" (2000: 281).

There are further important differences in media uses between the two countries. Sweden has the highest daily newspaper readership among European countries (75 percent), and while a third of Americans say they read a daily paper, almost half of Europeans say they do so (figures for 1998 and 1999 surveys, Norris 2000: 79).[14] Watching the news on television "tended to be more uniform across postindustrial societies" Norris continues, but the "lowest viewing rates were found in France and the United States," with Sweden higher than the United States for both daily listening to radio news and watching television news (2000: 79–80 and table 4.2 on p. 80). Thus Norris groups Sweden among the "newspaper centric societies . . . characterized by extensive reading of the press and relatively little attention to TV entertainment" and the United States among the "television-centric societies" which "feature intensive use of TV entertainment and low newspaper circulation" (2000: 85).

The difference between a newspaper-centric and a television-centric society must immediately be placed alongside another major difference that bears on politics (and thus political advertising), campaign finance. Parties have been publicly financed in Sweden since 1965, and taxes are therefore the main source of income of all the parties, including for their increasingly capital-intensive electoral media campaigns (Åsard and Bennett 1997: 157). In Sweden, election campaigns are also funded by the state, as opposed to parties and candidates having to raise money in the United States (with the partial exception of presidential campaigns). According to Norris, in recent U.S. presidential campaigns, 60 percent of all spending went to television and radio advertising (2000: 152). It is difficult to compare the two countries directly, but Åsard and Bennett note that campaign spending on a single California Senate election cost more than twice as much as a national election in Sweden for all parties and all offices in the same year (that is, 1994) (1997: 12, though California's population is more than three times the size of Sweden). Still, the United States also has laws providing equal access to media that counteract the importance of campaign finance to a large extent.

In America, election campaigns became more oriented to the national level in the 1960s and 1970s. This nation-wide orientation of the news and of national election campaigns has remained in place even after the proliferation of television channels (Norris 2000: 170–71). Norris argues there has been a "shift from direct to mediated forms of campaigning" (2000: 156) and that the defining features of recent American (and British) election campaigns— she calls it the "postmodern campaign"—are "the professionalization of the campaign consultants, the fragmentation of the news media system, and the dealignment of the electorate" (that is, groups of voters being less tied to particular parties; 2000: 178). The "professionalization of the political consultant industry," in particular, Norris says, "has developed furthest in the United States . . . outside of America . . . [it] has been slower, mainly because parties have incorporated professionals within their own ranks" (2000: 146).[15]

Yet Åsard and Bennett note that political consultancy has also grown rapidly in Sweden, particularly since the 1980s, but this is centered more on parties rather than on interest group politics, which play a greater role in the United States. One reason for the rise of issue-based interest groups in both countries, but particularly in the United States, is that they have stepped into the space vacated by voter dealignment from parties (Åsard and Bennett 1997: 168). Again, these interest groups and their "machines" are not unique to the United States, but they are more advanced there than elsewhere. This also applies to the monitoring of public opinion, though both countries have seen more strategic uses of the media system, with political consultants and opinion polls measuring public responses in increasingly sophisticated and systematic ways.

In terms of content, analysts agree that there has been an increase in "soft news" (human interest stories rather than "hard" policy analysis) in the United States and in Sweden and other countries (for example, Norris 2000: 106, 109). And in Sweden, although there has been a great increase in the number of hours broadcast *overall* (with longer hours of transmission via more channels), the *proportion* of programming devoted to news has dropped on public television channels, "significantly in some of the Scandinavian countries" (Norris 2000: 106). And again, in the United States, the main change has been that news has become "balkanized" (Norris 2000: 101), no longer dominated by the three national networks as in the 1960s but distributed across the many new satellite and cable channels.

Do the media play a greater role in American politics than elsewhere? Norris says that in the United States, "the extreme fragmentation of authority among government institutions and the permeability of those institutions, combined with the weakness of parties as the 'glue' holding the political system together, may allow the news media (especially network TV news, the major newspapers, and a handful of policy-oriented monthlies) to play a much more powerful role in the policy-making process than in most other established democracies" (2000: 280). This is not to say that these several media outlets necessarily ensure a diversity of political views. As Page demonstrates with reference to several examples of major events, sometimes a few dozen professional communicators dominate the framing of a particular issue across *all* media outlets in the United States (1996: chapter 5).

But to address the role of the media in politics in more depth, we need a more systematic comparison between the two countries, and this can be done by reviewing two major arguments that have been put forward in relation to the two systems of mediated politics, one a direct comparison between Sweden and America made by Åsard and Bennett, and the other Hallin and Mancini's comparison between the "liberal" (Anglo-American) media systems as against "democratic-corporatist" ones best exemplified by the Nordic countries.[16]

To begin with Åsard and Bennett, one contrast between the two systems that might be expected in view of the more television-centric and commercial nature of the American system is that American audiences have a shorter or more limited attention span in relation to major political issues. In Sweden, on the other hand, where political debate is more driven by parties, newspapers, and public broadcasting, it is possible to sustain debates and agendas on a longer-term basis. In this respect, Åsard and Bennett contend that Sweden has in recent decades moved closer to the United States, and this makes sense in light of the recent addition of commercial broadcasting. They also argue that it is still easier to inject new ideas into the political process in the comparatively "centralized and interconnected" Swedish political system than in the American one with its "complexity" and "decentralized relations" (1997: xvi).

Åsard and Bennett's argument is based on an examination of several case studies of major political debates in different periods in the two countries: the New Deal (United States) and the Folkhem ideal (Sweden) in the 1930s and tax reform in both countries in the 1980s and 1990s. Based on these comparisons, they conclude that although "the U.S. ideas market is more open and competitive—or at least it is less coherently regulated—than the Swedish one," yet "governing ideas dissipate faster and meet with stiffer opposition in a less regulated communication system than in a more regulated marketplace [of ideas]" (1997: 180, 181). This is a result of changes in both countries in the relations between politics and the media whereby politicians increasingly only aim to augment their short-term support via the media system, and the media system, in turn, reflects and contributes to this diminished attention span.

Put the other way around, it is difficult for new ideas to be introduced and for politicians to engage citizen-publics with new agendas under these conditions, and the difference between the United States and Sweden has diminished in this respect. Åsard and Bennett suggest that this short-termism is a feature of the 1980s and 1990s. To keep themselves in power, politicians' use of the media has moved away from building long-term public support for competing visions that are, for Åsard and Bennett, part of the fundamental political democratic process in society.

In Sweden, they argue that the long period of consensus in Swedish politics that characterized most of the twentieth century broke down in the 1980s and 1990s. This consensus, consisting of a stable party system in which a dominant party (the Social Democrats) had to compromise and govern with other parties, meant that a wide range of interest groups could be incorporated into the political process. With the consensus around this vision broken down, and with new parties entering and fragmenting the political system, they believe that there is now less inclusion of interest groups and more agenda setting by politicians themselves.

Above all, however, Åsard and Bennett say that the media are to blame for short-termism in both countries. What they say about the visions of politicians becoming less engaging and the public therefore becoming more apathetic is not the central concern here. For our purposes, we can also leave their argument about the (somewhat) more diffuse state of American political communication to one side for the moment—although their (counterintuitive) argument that the Swedish "marketplace of ideas" is no less vigorous for being more regulated is worth keeping in mind. What is important is that they lend support to the more general point being argued here; that the relations between the media system, political institutions, and the public have become increasingly mediated; in other words, their account shows how ever *more* mediated politics in the two countries has not led to a more dynamic public sphere, but rather the reverse.

With this, we can turn to Hallin and Mancini, who also take a long-term historical view to compare media in different countries. Unlike Åsard and Bennett, however, who focus on the "marketplace of political ideas" and how it is influenced by the media, Hallin and Mancini are more interested in different media "systems": "The Liberal Model," of which America is the prime example, "is characterized by a relative dominance of market mechanisms and of commercial media," whereas Sweden is an example of "the Democratic Corporatist Model," which is characterized "by a historical coexistence of commercial media and media tied to organized social and political groups, and by a relatively active but legally limited role of the state" (Hallin and Mancini 2004: 11). One of their main arguments is that "national differentiation of media systems is clearly diminishing" (2004: 13), although they also argue that differences remain in place.

Hallin and Mancini want to avoid making a contrast purely between a more commercial versus a public service system, which is how the contrast between Europe and America is often seen. In America, for example, there has always been a strong ethos of a "neutral" journalism, and Hallin and Mancini argue that the "evidence suggests that there is no necessary connection between commercialization of media and neutral professionalism" (2004: 286). In fact, as mentioned earlier, Sweden can be said to have partly imported this journalistic ethos from America. The mirror image, the idea of a toothless media under a publicly subsidized system, however, is equally baseless. Hallin and Mancini argue that the media have taken no less of a "watchdog" role in the light of being subsidized by the state, especially with the rise of critical journalism in Sweden during the 1970s (2004: 162–63).

The contrast should thus not be made in terms of commercial versus public media, but rather in terms of the relation between the state, the media as a separate institution, and social groups: In the democratic corporatist model, the state is regarded as in partnership with or bargaining with social groups, and this is echoed in "a notion of journalism as a public trust" (Hallin and Mancini 2004: 190, 192). Democratic corporatist societies are "characterized by an ideology of collective responsibility for the welfare and participation of all groups and citizens. This is reflected in the media field by a strong consensus that the state must play a positive role as the guarantor of equal opportunities of communication for the organized social voices in pursuit of the 'common good'" (Hallin and Mancini 2004: 197). In the liberal model, in contrast, "the role of organized social groups . . . is often seen in negative terms, as elevating 'special interests' over the 'common good' . . . the role of the media tends to be seen less in terms of representation of social groups and ideological diversity than in terms of providing information to citizen-consumers and in terms of the press as a 'watchdog' of government" (Hallin and Mancini 2004: 298–99).[17] This chimes well with what has been said so far about the Swedish system being aimed at the public good as against the more adversarial American system.

The point, again going against conventional wisdom, is therefore not *just* a unidirectional increasing drift toward greater commercialism. Hallin and Mancini note that in addition to commercialization, "media system change has played an independent causal role," including "the rise of television, the development of 'critical professionalism,' and media markets" (Hallin and Mancini 2004: 295). Thus they point out that "commercialization contributes to a shift in the power between the media and political institutions, with the media themselves becoming increasingly central in setting the agenda of political communication" (Hallin and Mancini 2004: 278). Media institutions are not just becoming differentiated from political groups, they also play an increasingly central role in societies in which people are no longer connected by direct ties. And commercialization is partly a response to this need for denser communication networks in dispersed societies; in other words, a dedifferentiation from the economic system (Hallin and Mancini 2004: 288, 291).

Still, despite these greater needs for commercial communication, and although liberalization and deregulation have made for a weakened connection with the state, Sweden's adherence to the democratic-corporatist model—with subsidies, regulation of political communication, and public service broadcasting—remains in place, albeit in diminished form (Hallin and Mancini 2004: 291–92). Nevertheless, Hallin and Mancini acknowledge that "commercialization . . . changes the function of journalism . . . undercutting the plurality of media systems . . . and encouraging its replacement by a global set of media practices" (2004: 277). This process, which originated at least as far back as the 1960s, has intensified since the 1980s, they argue (Hallin and Mancini 2004: 277).

Commercialization therefore enhances the autonomy of media institutions in relation to the state, but it is not the only trend. The multiplication of media channels has affected all societies in the direction of greater pluralism (Hallin and Mancini 2004: 295). And the concentration of media ownership for the press, although it has continued, has been slowed even if it has not been reversed, by means of state subsidies (Hallin and Mancini 2004: 162), and in the United States by regulatory and legal means. Furthermore, even with the lines between entertainment and information becoming increasingly blurred, neutral professionalism is "likely, not to disappear, but to find itself reduced to one genre among many" (Hallin and Mancini 2004: 287).

They conclude that "differences among" media systems, "and in general the degree of variation among nation-states, has diminished substantially over time . . . the Liberal Model has clearly become increasingly dominant . . . across much of the world . . . its structures, practices and values displacing, to a substantial degree, those of other media systems" (Hallin and Mancini 2004: 251). This includes the "principles of 'objectivity' and political neutrality" becoming "increasingly dominant" in journalism; the spread of an American style of "dramatized, personalized, and popularized . . . broadcast journalism," and political communication moving "away from party-centred patterns rooted in the same organized groups as the old newspaper system, toward media-centred patterns that involve marketing parties and their leaders to a mass of individualized consumers" (Hallin and Mancini 2004: 252).

In short, Hallin and Mancini describe a complex process: *some* convergence on the Anglo-American liberal model, but with the democratic-corporatist model also remaining in place. And the convergence is neither a straightforward dedifferentiation (politics losing out to commercialism), nor simply the greater differentiation or separation of commerce from politics. Instead, the media have become more commercial but *also* more autonomous—but *within* this institution there has been more commercialism in addition to proliferation, both of which contribute to maintaining a neutral professional style of professionalism as one option in both systems. The main shift, however, has been in the relation between state/media and society, and this is where commercialization has to some extent eroded the link between organized social groups and media acting on behalf of the common good.

The conclusions of both studies and the common ground between them can be put in terms of the theoretical positions about the media and politics outlined earlier: there has been some convergence on the liberal model, but the two models also remain different; and although there has been greater commercialization in Sweden in recent decades, both media systems have also become more autonomous over time. This leaves media institutions in a state of being suspended between increasing commercial pressures and their own more autonomous and diversified role. The media are therefore not, contra Habermas, being submerged by commercial interests and technicist politics (dedifferentiation), nor, as for Parsons, are they simply an independent feedback mechanism (differentiation) that allows the public's values to become integrated into the political sphere. Instead, mediated politics has a dynamic of its own, becoming more commercialized alongside developing a momentum as a growing autonomous system that is more and more routinely used in a strategic way by media and political elites, which includes gauging public opinion, as well as being routinely engaged by civil society's movements and organizations.

A shorthand for this process could be to say that mediated politics has become increasingly "managed" by all three groups between which they are suspended—political and media elites on one side, and citizen-publics on the other. The interface where they meet, the public sphere of mediated politics, is partly characterized by conflict—the media's "watchdog" role, politicians getting their agendas across, civil society seeking a conduit for its demands. But mediated politics consists not only of conflict or tension, but also of mutual accommodation between the three actors. It consists of a process whereby the input of citizen-publics or of civil society is increasingly gauged and proactively managed and has taken on a dynamic of its own. As Åsard and Bennett point out, targeting the media has become an increasingly strategic and scientific process (1997: 16).

This growing autonomy must be located in an overall social context. The main point established by Åsard and Bennett *and* by Hallin and Mancini concerns the difference in the relationship between state, media system, and society: the American state is more penetrable by the media, but in Sweden, the role of the media in promoting the common good is undermined by their increasingly plural and commercial role—even if they still play a more coherent role than in the American system. But this difference and partial convergence must also fit into the context of the pressures "from below": the system of mediated politics confronts or faces citizen-publics in different ways; fitting into the penetrable gap between state and society in the United States, as against incorporating citizen-publics into the common good in Sweden.

This is the broader "social shaping" of the large technological media system,[18] the struggle by civil society (citizen-publics) for greater penetration of elite democracy. This struggle—conflict—is ongoing and waxes and wanes historically (Collins 1992).[19] Overall, therefore, it may be easy to agree with Norris's view that "postindustrial societies have seen diversification in the channels, levels, and formats of political communication that have broadened the scope for news and the audience for news, at both highbrow and popular levels" (2000: 311). What this view ignores, however, are the constraints of the media system: Norris's view is based on a non-zero-sum conception of political communication, a conception where political communication offers as many choices as users can take advantage of. And although the scope of political communication has expanded over the course of history, we shall see that there are clearly limits.

It has been argued here that the large technological—media—system encompasses the *range* of mediated politics. One limit then is that across the whole of mediated politics, there is a limited window on reality both on the side of the content that is produced (by media elites) and on the side of the content consumed by audiences (citizen-publics). As both Gans and Carey have argued, the proliferation of media does not entail boundless diversity; the attention space and what fills it are finite.[20] As Carey puts it pithily, "reality is . . . a scarce resource" (1989: 87). Similarly, Gans has analyzed how the sources of news, the social background of media professionals, and what is regarded as main target audience—all serve to produce a particular outlook on the world, which he calls the "national newshole" ([1979] 2004: 319).[21]

These points need to be combined with the fact that this "newshole" is the *only* commonly shared reality in society. As Collins points out, "the mass media are the only place where there is a recurrent focus of attention shared by anything close to a majority of the society" (2004: 279). And even this focus of attention is fleeting in the American context where "diffused social conflict," as Hall and Lindholm point out, makes it "very difficult indeed for the society to concentrate on a single set of of issues for any extended period of time" (1999: 76). As we have seen, Åsard and Bennett say that the same has come to apply to Sweden. And Hallin and Mancini note that "in general, the development of the media in the twentieth century led to an increased flow of culture and information across group boundaries, reducing dependence of citizens on exclusive sources within their particular subcommunities" (2004: 271). But while there may no longer be "exclusive sources," the sources that the large technological media system provides at any given time are limited by the "affordances" of the dominant media and by the limited attention space that they make available.[22]

This argument can be taken further by reference to Luhmann's claim that "whatever we know about our society, or indeed about the world in which we live, we know through the mass media" (2000: 1). This claim exaggerates and is based on his exceedingly broad definition of communication.[23] After all, we also learn about the world without *media* in the conventional sense of the term. His main argument, however, is that what we know through the media has little effect—except to keep us abreast with the increasing complexity of society via a feedback mechanism between public opinion and political decision making (2000: esp. 82 and 105). Since this feedback is not, as for Parsons, a transmission of society's inputs into the political process but rather simply part of an ongoing conversation that society has with itself, Luhmann is making the point that just because there is *more* mediation of politics, this does not necessarily entail greater pluralism and deeper democracy, but it may simply be a complexification—not of the political, but of the social process as a whole.[24]

Luhmann's ideas can be contrasted with two functions that are conventionally ascribed to media mentioned earlier, that they create and sustain a belief in the legitimacy of coercive power (the Weberian conflict view), and that they may contribute toward a sense of greater participation in society (the functionalist, modernization view) (Thompson 1995: 15, 190). In recent times, it is necessary to add to this, as Thompson suggests, that "struggles for visibility" have become increasingly important (1995: 247). But what if the legitimating function has become so taken for granted as to be invisible, and the participatory function ever more "managed" and subject to being used strategically? This would support Luhmann's neofunctionalist and constructivist perspective, which holds that the media do not "socialize" people into a particular view of the world or "increase" their knowledge or enhance their capabilities as critical citizens. Instead, the media mainly serve to perpetuate themselves (2000: 98) and constitute an ongoing "irritation" to the political process—that is, an input that rechannels the social system rather than contributing directly to political decision making.

With this, we can return to one of the central questions of this chapter: do mediated politics "deepen" popular democracy? From the point of view of the material that has been discussed here, with its focus on technology, systems, and a long-term historical perspective, there are two ways this could happen: one is a more penetrable form of mediated politics with greater input into the limited attention space, and the other is more scope for "pressures from below" to become incorporated into political process. Only providing greater access could serve the latter role, and only an unfreezing of the locked-in relationship between the media system and the public could open up the former. In both cases, new technology could play a decisive role, though it would need to fit into the existing media system. And a thaw or reconfiguration of the media system on the social (as opposed to the technological) side would be much harder to achieve because of the congealed nature of this large technological system.

Enter: The Internet

This brings us to a brief consideration of the role of the Internet. What effect can this new medium have on political participation and on the diversity of—and access to—political communication? Bennett (2003) and Bimber (2003) argue that one of the novel capabilities of the Internet for political change is that there are fewer gatekeepers (or lower entry costs) and being able to share a common frame of reference with distant others.[25] But Bimber, who puts the Internet in a larger historical context, is also more cautious. With recent increases in information abundance, he says, "information is becoming less politically institutionalized" (2003: 229). He notes that "the number of elites and potentially viable mobilizers appears to be increasing, and competition for political attention [is] growing more aggressive, against a background of largely unchanged habits of political knowledge and learning" (Bimber 2003: 230). At the same time, traditional mass media—broadcasting—will be more effective at reaching large audiences, and they are more expensive and dominate the attention space, especially in those arenas where decisions are made by large majorities (Bimber 2003: 105–6).

As we have seen, the balance between the influence of newspapers and television differs in Sweden and the United States. Interestingly, therefore, Norris suggests that Internet users "are more similar" demographically "to newspaper readers than to television viewers" (2000: 129). At the same time it seems that the Internet is not replacing other news or information sources: Bimber says "there is scant evidence for a substitution effect . . . internet use is additive, and in a general way, the media-rich do seem to get richer" (Bimber 2003: 217). In any case, the uses of the Internet for obtaining information already show signs of stabilizing, at least in Sweden (Hadenius and Weibull 2003: 264).

These points reinforce the argument that has been made here: the Internet will complement and add to (rather than superseding and replacing) the other channels (print and broadcasting), and its contribution will need to fit within the existing technological system and how this system, in turn, fits into the political sociology of the two countries. The Internet adds to the diversity of media, and so to greater pluralism, but it is not so much the additional content as the technological capability of making a difference to the dominant one-way mass media described that will, as Bimber and Bennett also note, potentially make the key difference. Bimber's "main thesis . . . is that technological change in the contemporary period should contribute toward information abundance, which in turn contributes toward postbureaucratic"—read, less top-down—"forms of politics" (2003: 21). But he acknowledges that so far, this is mainly a potential that may or may not come to be realized in the future. And this, in turn, will depend on whether the Internet becomes a major modality within the larger system of mediated politics.

Converging and Congealing Systems and the Increasing Technological Mediation of Politics

The mass media provide a limited interface between media and political elites and citizen-publics, and although information and communication technologies have proliferated, they are constrained—by the congealing of the technological system on one side, and by an increasingly "managed" apparatus that mediates between political and media elites and citizen-publics on the other. This constraint is also imposed by a limited attention space (except perhaps at the margin where new technology provides an additional input, which must nevertheless fit into the existing system) and the historically conditioned patterns of dominant media uses that audiences have developed and that have been charted here.

The more varied and more extensive mediation of politics that has come to characterize Sweden and the United States has been a unilinear process common throughout advanced societies. This is one form of convergence based on the imperatives of technological development and of the development of technologies into a large system.[26] To this convergence can be added the convergence described by Hallin and Mancini toward the Anglo-American system with more market mechanisms and commercialized media. But these two forms of convergence—the one technological, the other due to "globalization," also allow us to pinpoint what is *not* so far converging; namely, the way the media system is embedded in two different political cultures and how these cultures continue to reinforce distinctive paths of political development.

Nevertheless, Hallin and Mancini underestimate convergence: it is not just media systems in the sense of governance of the media which, as they argue, have converged somewhat. It is also the cumulative effect of a number of technologies and their changing shapes that add up to similar large technological systems for mediating politics. These systems, again, have retained some of their differences (newspaper-centric versus television-centric, a system directed at the common good versus a more adversarial system), but the overall effects of the two technological media systems surely outweigh these differences—an extension of mediated politics in scale and scope, constrained by a limited newshole and limited attention space, combined with the dominance of certain media channels.

The public sphere is partly a conduit, but it has also become a political instrument—and thus it "cages" those who make use of it as well as providing a vehicle for them.[27] As we have seen, the process whereby political ideas get put onto the agenda, from above and from below, has become more mediated. New media have become added over time, but they have complemented rather than substituted for existing ones. From a historical point of view, the media (print, radio, television) have also succeeded each other as dominant modalities. Again, this succession of dominant modalities, and the shifting balance between them, is common to both countries (and elsewhere), but it is easily overlooked in the emphasis on the proliferation of media.

Changes in the two large technological media systems on the social side are therefore only one part. The technological side is equally important and has converged even more strongly: the *uses* of technology have become more similar, toward the use of a greater variety of media, but also toward the dominance of television, with all that this implies for the "scarce" reality and attention space of political communication. The key point about these uses of technology, however, is that they have become congealed around particular patterns and in this way provide a limited arena for the public sphere.

In short, it is possible to pinpoint the social shaping of the system: the differences in the political cultures that shape mediated politics and some convergence toward a common system. But this social shaping of the system needs to be put in a larger context of political and social change in the two countries: the penetrability of the American state and the diffuse role of the media, and social input constrained within the bounds of integrative consensus in the Swedish case. This is the larger social context that mediated politics confronts. But the system also has its own social and technological momentum: social, because the system has become more managed and allows limited scope to incorporate pressure from below; technological and social combined, because certain modalities have come to dominate mediated politics. And technological shaping, because the technologies of which the system consists have become congealed into a particular configuration common to both countries and thus provide a limited interface through which citizen-publics and elites communicate via the public sphere of increasingly mediated politics.

6 The Consumption of Technology in Everyday Life

M OST WRITING ON TECHNOLOGY UNTIL RECENTLY HAS FOCUSED ON economic growth and on production to the neglect of consumption.[1] One reason for the recent interest in consumption is that life outside of work—leisure and domestic activities—have become more prominent in the social sciences. Another reason is that the notion of culture has been widened: culture is no longer just ideas but also material artifacts, it is not just the content of media but also the means of dissemination, and it is no longer mainly cultural producers but also their audiences. Moreover, it has been argued that users don't just passively consume technology, but actively transform it. I will argue here that much is to be gained from the focus on consumption and everyday life, but, in line with the kind of technological determinism put forward throughout this book, I will reject the idea we should focus on how users shape or transform technologies.

This chapter will examine the consumption of technology in everyday life by considering three major technologies—car, telephone, and television. The argument, which extends the case for technological determinism put forward here and goes beyond the social shaping and social constructivist views, is that these technologies have had the uniform effect of diversifying leisure and sociable activities in developed societies. To make this argument, again Sweden and America, two countries for which detailed evidence is available for different periods of time, in this case also including several studies of "typical" small towns, will be used for a long-term historical comparison. In addition to synthesizing this evidence, this chapter will develop some neofunctionalist ideas about technology and culture (and extend the Weberian ideas that have already been put forward) to develop a model of how the consumption of technology in everyday life simultaneously homogenizes social life and makes it more diversified.

The point of departure for this chapter is Edgerton's (1998) argument (taken further in Edgerton 2006) that there has been too much emphasis on innovation and production, and not enough on the everyday *uses* of technology. Some recent writings about the consumption of technologies in everyday life have gone some way toward redressing this imbalance, but this shift has not had a clear focus. The easiest way to recognize this is to ask: what is the study of everyday life and consumption *for?* The answer that I will give goes against the grain of much contemporary writing on this topic, which has stressed that the consumption and uses of technology are shaped by different contexts and hence that it is impossible to generalize about their effects. I will argue that if we analyze the role of technology in everyday life, we see that technology has had rather uniform effects.

The argument goes further than this: the empirical study of everyday life and the consumption of technologies has so far concentrated on the microlevel of social change, and thus remained tied to particular periods and places. My argument, in contrast, is that these microlevel changes add up to—again, more uniform—macrolevel changes, and once we identify what these are, we will be in a better position to identify the distinctive role of technology in contemporary culture and society. This link between micro and macro, or between local and wider changes, has been ignored in writings on consumption and technology. Put differently, once the changes resulting for the role of technology in everyday life have been aggregated, we can separate out the effects of technology from the changes in the other spheres of life: politics, economics, and culture (*minus* the effects of technology). This will also allow us to tackle the vexed problem that was mentioned in Chapter 1 concerning the relationship between culture (which is often treated either in its particularistic local form or in a more abstract sense globally) and technology (which can be seen as global in the different sense of becoming diffused throughout modern society).

Finally, we will be able to go beyond two extremes in the study of technology and society, which can be represented here by Giddens and Fischer. Giddens (1990: esp.17–29, 137–149)—and following him Castells (2000)—has argued that modern individuals are becoming disembedded from their immediate spatiotemporal and social context and that we are increasingly participating in a more global social context. For Giddens, this change can to a large extent be attributed to information, communication, and transportation technologies. Fischer (1992), on the other hand, has argued that users shape technologies rather than the other way around, and from a social network perspective, he claims that (in his case the telephone) reinforces rather than disrupts existing social ties. To anticipate, Fischer's is the more difficult "extreme" position to go beyond, but I will argue that while it may be the case that users *initially* shape the *uses* of technology, the consumption of technologies in everyday life, once shaped, becomes an established part of social life, and then the question becomes how technology has distinctively shaped our culture (culture in this case rather than economics or politics because consumption in relation to everyday life mainly belongs to the cultural sphere), and how this cultural change relates to social change as a whole.

The use of the study of the consumption of technologies in everyday life then is that (1) it documents the distinctiveness of modern culture in comparative-historical perspective, (2) it allows us to pinpoint the role of technology in social change, and (3) it can help us to understand the impact and potential impact of ongoing and foreseeable technological changes—say, with the further diffusion of technologies or the introduction of new ones.

One final preliminary point is necessary: in this chapter, I will focus on how technologies become embedded in everyday social life, which means looking at consumption and end uses. But it is important to mention that there is in addition an intermediate stage, the "mediation" of consumption, a set of practices or institutions between production and consumption that have also recently moved into the foreground. These mediating institutions include consumer and marketing associations, for example, and the various ways in which consumers are taught and learn how to use new technologies. I would like to suggest that the main mediating institutions that one would need to include in the case of the three technologies in this chapter are "large technological systems" (Hughes 1987) described in Chapters 3 and 5, not on the side of production but of consumption—telecommunications providers, electronics vendors, highway and car maintenance services, and so forth. And again, as we saw earlier, one of the features of twentieth-century technologies is that they have become "systemic," and this applies particularly to the large technological systems that mediate consumption. Many of the important technologies in modern everyday life depend on these large technological systems which, once they have become "ossified," become taken for granted. From the perspective of the user or consumer then (unlike the perspective of the historian or social scientist), these systems are of little interest since their importance has faded into the background.[2]

To assess the changing roles of the three technologies, it will be useful to compare the roles (or the lack of them, or their precursors) at three points in both Sweden and the United States: before industrialization, before World War II, and today—or, translated into our three technologies, first none had it, then some had it, now (almost) all have it. Apart from yielding a rough before, during, and after divide for comparison, there is another good reason for looking at three slices in time: technology, unlike other social institutions (but like science), is both cumulative and rapidly introduces new ingredients into modern social life. This point about the role of science and technology in history, again, goes against the grain of much of the study of technology in society, from social shaping to social construction, which overlook these cumulative effects.

The topic of technology in everyday life is potentially vast, so I will concentrate here on three technologies that have undoubtedly had a major impact and that are widespread (and on which there has been much research). Put differently, the three that I have chosen are highly visible technologies (for some less visible ones, see Braun 1993). And I will concentrate on their consumption or uses.[3] But while the differences in the political and economic systems between Sweden and the United States have persisted, the argument will be that the uses of the three technologies in the two countries have increasingly converged. The main point of concentrating on these two countries is to tie the observations about the effects of technologies closely to particular times and places and thus to act as a brake on speculation. Yet in doing this, we will find that many of the observations below could just as well apply elsewhere.

The Cultural Significance of Everyday Uses of Technologies

To recapitulate briefly, technological advance is a process of "interlocking of refining and manipulating," refining being linked to scientific advance and rationalization, and manipulating being what technology *does* in relation to the natural or social environment. Thus technological artifacts, which always also consist of physical hardware, are continually being modified in order to enhance our mastery of the world. A more fine-grained distinction within technology could be made here between things, objects, artifacts, devices, and machines (see Braun 1993: esp. 41), but for our purposes, it will suffice to say that our three technologies are all artifacts. In modern society, technology therefore "disenchants" the world, constantly extending the impersonality of the external conditions of life to new areas.

Technology, because of the disputes mentioned earlier, and because of the emotive charge of the word, has had to be given a precise definition. Some of the other concepts that will be used can be defined in a more pragmatic way to delimit the subject matter. Thus, by consumption I will mean the area outside of work or outside of production, and thus free or noncommitted time for leisure or sociable activity. From the point of view of the social sciences, everyday life *is* culture in the sense that this notion captures the parts of social life that are outside the economic and political spheres of life. But in focusing on the consumption side of everyday life we encounter a seeming contradiction: that much of the nonwork part of the everyday seems to be aimed at getting away from everyday concerns in the sense of getting away from mundane or routine activities. This contradiction can be resolved by distinguishing between everyday in the main sense to be used here, as the arena of leisure and sociable activities, and the other sense of everyday (which will also occasionally be used, but clearly indicated as such), which involves getting away from everyday routines.

What about culture? Culture can, again, be used in a pragmatic sense as a way of life or the sphere of society outside of the economic and political spheres. For the purposes of this chapter, and since we are dealing with consumption, we can further narrow this down to the nonwork, private or household, and everyday realm. Following the earlier discussion in Chapter 1, culture then needs to be subdivided further into the area that is shaped by technology and the area that is not. Technology thus *becomes* culture insofar as it is translated into everyday life. But, on the definitions used here, science and technology are also separate from culture. How technological change relates to cultural and social change is then an empirical question, or in the case of the topic of this chapter, a question of how the everyday microchanges brought about by technology add up to larger, macrocultural and social changes. The topic of this chapter can thus be summarized—or visualized—as the intersection of consumption, technology, and everyday life. (And the image can be made more complex by

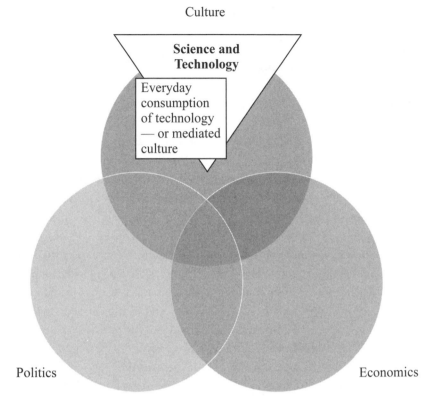

Culture

Science and Technology

Everyday consumption of technology — or mediated culture

Politics

Economics

FIGURE 6.1 The relation between the spheres of culture, politics, economics, and science and technology, including consumption or mediated culture.

locating consumption and everyday life as lying within the circle of culture in a picture of society as consisting of the circles of the spheres of culture/science and technology, politics, and economics; see Figure 6.1).[4]

But the translation of technology into culture will depend on the technologies in question. What their cultural significance "adds up to" in comparative-historical perspective therefore comes at the end of the investigation rather than at the outset.

Television, Cars, and Telephones in Everyday Life in Sweden and America

A good place to start comparing Sweden and America are two detailed sociological studies that cover everyday life, *Middletown* (Lynd and Lynd, 1929) and *Medelby* (Allwood and Ranemark, 1943). Both present material about daily life in two similar settings over a half a century ago, and both also make comparisons with the situation in the two places before the turn of the previous century—in other words, before the onset of industrialization (and for Middletown, though not Medelby, there are follow-up studies that take the study up into more recent times). The fieldwork, surveys, and questionnaires for Middletown were carried out in 1924–25 and for Medelby in 1941, and the latter study was deliberately modeled on the former. Both places were selected to be "typical": neither too large or too small, neither too advanced in industrialization nor too much lacking in it, and in the heartland of the country rather than being too remote or too close to a metropolitan center. With their size and location, both had recently been subject to a strong influx of industry and to the arrival of rail and transport connections.

· · ·

Compared to half a century earlier, a noticeable change—if we adopt a somewhat detached perspective for a moment—is that the role of the priest and the school teacher as primary mediators of culture had been diminished by the new media and that the authority of the local elite had been at least partially displaced by that of a more national elite. This was to a large extent due to the—then new—information and communication technologies: mass-circulation printed material and radio.

Religious reading material, which had contributed by far the largest share of reading before the twentieth century in both places, had become only one source of reading material among many. By the time of the two studies, almost all of the Medelby and Middletown households bought at least one daily newspaper. The main line of stratification in this respect is the one that has persisted to the present day (see Collins 1975, and Holt 1998: 12–13 in relation to consumption), between a "cosmopolitan" elite readership of a national newspaper and the more "local" newspaper readership. It should be added that in terms of content, politics in the newspaper, which is often considered to be one of the most important functions of newspaper readership, is of minor interest among the two populations in comparison with the entertainment or leisure part of the paper—local events and shopping opportunities, advertising, cartoons, sports, and the like.

Apart from daily newspapers, Middletowners and Medelbyans have begun to consume a wide variety of weekly and other magazines that cater to diverse interests—religion, adventure, hobbies, romance, and so on. What most of the content of these reading materials has in common is that it takes the readers away from everyday concerns—work and the immediate world around them—and into the world of noneveryday hobbies, places to visit, things to consume, fun, and adventure. The radio initially competes with the newspaper for delivering news and entertainment, but after a period of enthusiasm, it eventually comes to be a complement to print and later broadcast media, and increasingly also becomes a secondary or background activity. (Another complement, the cinema, has already become a popular pastime in Middletown, but there is as yet only a—very popular—tent cinema in Medelby at this time.)

Cars have now also put various sorts of leisure within reach. From the side of the infrastructure, the recent arrival of automobility has caused some conflict over the resources for building and maintaining roads. But from the user side, and bearing in mind that a lot of the early use of cars was for recreational purposes, it is also noticeable how quickly the car has become adopted—and adapted to—rather than causing conflict, despite the considerable new financial burden that this put on the users.[5] The main uses of the car (and to some extent of trains and buses) are the increasing number of leisure and shopping trips that an ever wider part of the Medelby and Middletown populations undertake.

Middletown and Medelby thus bear out the point that print, communication, and transport technologies are often misperceived, as Fischer (1992), Nye (1997), and others have emphasized: it is not primarily that new print media provide users with the (instrumental) means of access to a wide world of knowledge, or that cars and other forms of transport provide the means to get to places more quickly. Instead, reading and driving become leisure activities in themselves. And they do so at the same time that leisure becomes a separate (set apart from the rest of the day and on weekends, and located in the realm of private life), noneveryday (getting away from routine), and regular—and in this sense everyday—pursuit, and the consumption of technology as a means to pursue leisure also becomes a widespread, regular and in this sense everyday part of life.

If these are the effects of the new technologies, what do the people in the two towns themselves think about the influx of new technologies and their influence on social life? As ever, they are initially concerned mainly about the impact on "morality" and on the disruption of "tradition," a concern that eventually subsides. This fear is a common pattern with new technologies that threaten to affect social life, as is its opposite, technoenthusiasm, and these fears and hopes are perhaps best viewed from a detached perspective as constant accompaniments to the most visible new technologies in modern societies.

In a similar vein, it is possible to regard both studies themselves from a critical perspective, as expressing the biases about technology of the researchers that carried them out, or of the time in which they were written (see Caccamo 2000 and Thörnqvist 2000). But if we regard them in this way, we get an interesting result that sheds light on the relation between technology and society: in relation to the impact of technology one can, for example, read between the lines of the avowed social scientific objectivity of Middletown a romantic criticism of technology as corroding the cohesion of American society. Similarly, in Medelby, one can detect a typically Swedish concern that the positive or modernizing beneficial effects of the new technology should be spread more evenly among the population and that the state should step in to remedy this shortcoming (this is particularly clear in an editorial by the co-author of the study in a major Swedish daily newspaper that appeared just before the publication of the book; see Allwood 1942). In short, we get the impression of characteristically national concerns.

With the benefit of hindsight and from a comparative perspective, however, we can also put these predispositions or biases themselves into context, inasmuch as the argument presented here suggests that the reasons for the criticisms expressed in the two studies are unwarranted; that is, there is no such "corrosive" effect in Middletown in the end, nor an "uneven" spread of technology in Medelby over the longer term. All the same, these biases do not seem to me to affect the evidence presented in the two studies or the thrust of the descriptions of the two places at the time, and this evidence and these descriptions can, as we shall see later, be used to draw different conclusions.

The overall effect of technology on cultural change in both cases is therefore that, in the course of industrialization, an extended range of technological means has been used to pursue more varied forms of leisure. Perhaps the most striking feature in both cases is how the growth of leisure time—or more accurately the emergence of a regular space and time set aside for leisure—was filled up with new information, communication, and transportation technologies. Put differently, there is a pluralization of leisure and sociable activities inasmuch as they become more diverse (and the technologies with which they are pursued more widespread), they occupy more time and are spatially more wide-ranging, and these differences in the ways of life are technologically mediated.

Bearing this new situation in the midst of "industrialization" in mind, we can now fast forward to more recent times.

To do this, we can begin with Erickson's (1997) ethnographic study of two small towns similar to Medelby and Middletown in contemporary Sweden and America. Although her study does not aim at a comprehensive sociological portrait like the two earlier studies—she is mainly concerned with consumption, energy, and the environment—the study nevertheless covers much of the same territory as the earlier studies and of this book. Erickson lived in the two towns, Munka Ljungby and Foley, during two periods in the early 1980s and again in the 1990s. Like the two earlier studies, she used a combination of participant observation, questionnaires, and surveys, though she is an anthropologist. And again, the two towns were selected to be typical in terms of size, economic makeup, and distance from metropolitan centers.

A number of points in Erickson's study are worth mentioning. The first is that the stereotypes of the conserving Swedes and the wasteful Americans—which were common among her informants themselves—were only partly borne out (1997: 3–5). What she finds striking instead is the gap that can be found in both communities between concerns about excessive consumption, energy wastefulness, and environmental deterioration on the one hand—and personal behavior, which is largely not connected to these larger concerns, on the other. Another noteworthy feature is the convergence between the two communities over the period that Erickson covers. As she says at one stage, "Sweden increasingly resembles America materially" (1997: 8). A further similarity that she noticed is a common rhythm in certain attitudes—for example, the increases in concern about energy after the oil crisis in the 1970s, which then wanes in both places in the 1980s and 1990s.

There are further interesting comparisons, but since Erickson's findings in relation to specific technologies and patterns of consumption fit in with the other contemporary studies of our three technologies that I will make use of, I will intersperse her material with these studies below. It should also be added, again, that we can use Erickson's results about consumption while ignoring the prescriptive part of her study—in this case, wishing to curb an excessive materialism and environmentally destructive practices by means of a renewed spiritualism and changed attitudes toward nature.

Before discussing the use of technologies further, it may be useful to say something about the general cultural similarities and differences in which the usage patterns that will be described have become embedded. The most important difference in relation to consumption and everyday life is perhaps the difference between two types of individualism.

In America, as Hall and Lindholm argue, a deeply pragmatic attitude combined with the idea of continual self-improvement has meant that Americans have a need to display the evidence of self-transformation in outward signs—accumulation or consumption being foremost among them (see also Nye 1998: esp. 182). In Hall and Lindholm's words, "the pervasive pragmatic modular approach to life permits Americans to . . . [visualize] the world around them as a machine that can be retooled, or taken apart and rebuilt, in order to achieve maximum efficiency . . . even the self is considered to be a kind of modular entity, capable of being reconfigured to fit into preferred life styles" (1999: 86). And: "each striving individual seeks to become 'all you can be' through ceaseless labor, accumulation, consumption, and display" (1990: 90).

Swedish individualism, in contrast, is oriented more, on the one hand, toward nature, and the peace and isolation that can be found there, and toward living up to the expectations of a communitarian society in which norms are highly transparent on the other (see also Frykman and Löfgren, 1987). The characterization by Orfali sums this up: "The dream of every Swede is essentially an individualistic one, expressed through the appreciation of the primitive solitude of the vast reaches of unspoilt nature" (1991: 443). "In Sweden, perhaps more than anywhere else, the private is exposed to public scrutiny. The communitarian, social democratic ethos involves an obsession with achieving total transparency in all social relations and aspects of social life" (1991: 418).[6] At the same time, in Sweden, consumerism has of course also become a means to express individualism (Löfgren 1995).

The different implications for consumption (for example, America is often regarded as the apogee of materialism and Sweden as the home of a strong environmental consciousness) and the similarities will be readily apparent.[7] These cultural differences may also have more specific consequences for the issues discussed here, especially energy consumption in relation to transportation. But the cultural characteristics of individualism that Swedes and Americans share may be more important than the differences. As Löfgren points out in his study of vacationing (especially in Sweden and America), "the credo of modernity is 'life can always be improved'" (1999: 268). This may be why studies like *Medelby* and *Middletown* show such similar patterns of consuming new technologies, or why Munka Ljungbyans and Foleyans share basic attitudes about consumption.

Note too that they share the same *contradictory* attitudes: Munka Ljung-byans and Foleyans are tired of materialism and excessive consumption, but among both populations, shopping has nevertheless become a more popular pastime over the past two or three decades, and the pattern of uses of increasing amounts of energy have resumed—once more energy-efficient technologies have been taken out of the equation—after a period of net savings and increased awareness in the wake of the energy crisis of the 1970s (Erickson 1997).

Put differently, there are cultural differences that shape the *forms* of consumerism in Sweden and America, but these should not be exaggerated. As Campbell (1987) has shown, there is a common cultural source for the constant striving for new experiences in modern societies, and this striving has become firmly institutionalized within a stratified status hierarchy. This hierarchy, in turn, means that consumption is not just an individual pursuit, but entails social emulation. And the social mechanisms for this emulation are well known: as Corrigan (1997: 171) argues, the aspiration toward an ever-receding horizon of status goals or goods should be regarded, not in terms of Simmel's "trickle-down" model, but rather, following McCracken (1988: 94), as "chase-and-flight"—that the ante is constantly being upped for high-status groups, which need to consume ever more in order to maintain their social superiority.

Thus, individually and socially, the culture of consumption expresses itself in a constant stream of new goods, especially consumer technologies. What Sweden, America, and other industrialized countries thus have in common is that they are "consumer cultures," a culture that has achieved a *stable* form, and that is at the same time continually *changing* because of the combination of new technologies and high economic growth (macro-) and new modes of experience and experience seeking (micro). As Mintz puts it, "the history of United States society, particularly since the end of World War Two . . . has been one of a long-term process of . . . the intensified ritualization of consumption" (1997: 198), and this could just as well be said for Sweden and other industrialized societies. With this, we can turn to a more detailed consideration of the everyday consumption of each of the three technologies in contemporary America and Sweden.

Television is still changing, but it has also become a stable part of everyday life. As Silverstone (1984: 97–103) has argued, the technology took the form that it has today in 1950's America, when the medium was "domesticated" in the suburbs, and used partly for information but mainly for entertainment. The television set took its pride of place in the living room, as it still does in

Sweden and America today, and it offered a type of entertainment that was suitable for—and acceptable to—a wide family audience.

In terms of content, as Meyrowitz has pointed out, the similarities in what people watch are more striking than the differences (1985: 79–80, compare Höijer 1998: 276). And it is only within this overall uniformity of how television has become embedded in everyday life that we can notice the main difference in how different groups use this technology, both in Sweden and in America: children and pensioners are the groups that deviate most from normal viewing patterns both in time spent and in content (Meyrowitz, 1985: 79–80; Höijer 1998: 262–65). We can also notice, as with cars and telephones, a general proliferation of apparatuses—multiple television sets and related devices.

But the most important feature of television is that it has become the single largest filler of the emerging niche of available leisure time, and, despite differences, Sweden continues to converge with American patterns in this respect. Like the advent of mass-printed material and radio one or two generations earlier, television has come to occupy a central place in everyday life and introduced diverse content into recreation. In this case, however, it is not content but the change in the use of time that is dramatic: as Robinson and Converse already noticed in the 1960s, television constituted the single largest change in the daily use of time, displacing a number of other activities since its arrival (Robinson and Converse, 1972). This trend has continued, though less dramatically, into the 1990s, such that television watching now occupies 40% (or 15 hours) of the weekly free time of adults in America (Robinson and Godbey, 1997: 125).

How have the everyday patterns of watching television changed over the course of time? Höijer (1998) has argued that in the Swedish case, there has been an important shift: in the early days of the 1960s and 1970s, when Swedish television was restricted to one and later two state-owned channels, television viewing took place in the context of the family gathered around the set at certain times and for certain programmes. These programmes were therefore often shared by the whole nation. Nowadays, with the addition of two additional broadcast channels, one of which has commercial advertising, plus video and DVD players, satellite and cable, as well as a much wider offering of programmes—especially American feature films and series and also entertainment shows—watching patterns are much more individualized, fragmented, and diffuse. This is also partly made possible by the fact that the majority of households now have more than one television set, as in the United

States. And although viewing hours per day have only increased somewhat over the past decades in Sweden (up by half an hour per day to two and half hours; Höijer 1998: 263) and a plateau has possibly been reached, it is clear that the audience is becoming more fragmented while the use of television as an "escape" or as relaxation has become more common and is now routine.

It may be that there are differences in television watching between Swedes and Americans, such that Munka Ljungbyans watch less and say (in Erickson's study) that this is a less popular pastime than do Foleyans (1997: 49–50, 119). But we should also note that Erickson found, like Höijer, that Swedish viewing patterns became more similar to American ones as the more diverse and entertainment-oriented (often American) offerings became more widely available in the 1980s and 1990s. The difference in viewing patterns that will have existed earlier because of the difference between the Swedish state-owned broadcast system and the American commercial one (as seen in the previous chapter) has thus become weakened, partly as a result of users' preferences. It is also interesting to note that, when asked about the relative importance of different consumer electronic devices in the home, Swedes rate the television set as being more important to them than do Americans (Venkatesh 1999).

With this, we can turn to cars. Two points stand out immediately in the longer-term changes brought about by car use: the first is that cars have changed the nature of leisure, making the family holiday by car a widespread middle-class institution. This is a major change from prior to the advent of the car, when extended family holidays away from home were restricted to the upper classes (Flink 1988: 169; Löfgren 1999: esp. 69, 90). This also goes against the widely held belief that that the automobile has led to isolation at the expense of sociability: on the contrary, in relation to leisure, the car has been used much more for recreation than is commonly believed, and in this respect the car has increased or enhanced sociability according to a number of writers (Fischer 1992: 253; Löfgren 1999: 63).[8]

Second, in the case of the car, unlike the telephone and television, the differences between Swedes and Americans make a difference. Nye points out (1998: 223) that half of the difference between European and American energy consumption can be attributed to transport, and if a substantial portion of transport is devoted to recreational or leisure uses, the impact of consumption on the environment is considerable. The lower energy consumption for transport does not mean, however, that Swedes think any less of the car as a vehicle for pleasure; as Hagman shows, Swedish advertising has played very much on this theme, and he also notes that Swedes drive cars with a higher average energy consumption than other Europeans (1998: 36).

If the car, in Fischer's words, more than anything "added to the sum total of social activity" (1992: 253), or, according to Nye, has mainly served to give expression to a "pre-existing penchant for mobility" (1998: 177) among Americans, then we nevertheless need to add to these views some characteristics that they leave out: one is that car driving is the second-most expensive item of household consumption (after housing itself, if that is counted). In Sweden in 1992, 14–20 percent of household income was spent on transport according to Polk (1997: 204) or doubling from 8 percent in 1950 to 16 percent in 1985 according to Kajser (1994: 198)—and similar estimates of up to 20 percent of household income can be found for Americans.

The second characteristic they leave out is that car travel is not far behind television in being a major daily activity, despite the fact that the time allocated for travel is restricted in a person's daily life. For example, in Sweden, where, as in America, car travel is dominant, the amount of time has not changed very much in recent decades—just over one hour per day, according to Vilhelmson (1999: 179), or ten hours per week in America according to Robinson and Godbey (1997: 117, though the two should not be strictly compared since different measurement techniques were used). What *has* changed, albeit slowly, during recent decades is the "activity space" (just over 50 kilometers per person per day in 1995 in Sweden), and also the number of cars per person (Vilhelmson, 1999: 178). This is part of a longer-term trend toward greater daily mobility in Western societies, and for our periods, it can be mentioned that Kajser estimates that for Sweden, daily travel by vehicle was .5 kilometers at the turn of the century, approximately 3 kilometers in the 1930s, 20 kilometers in the 1960s, and 40 kilometers in the 1990s. It should also be noted that "free time activities . . . dominated the total trip length . . . by almost 50% of the total" in Sweden (Vilhelmson 1999: 178), or again that, according to Robinson and Godbey's estimations, it is "free time" travel (and travel for child care) that shows a small increase in recent decades in America (1997: 117).[9]

Uses of the telephone present a different set of issues. Fischer's (1992) study of the telephone is one of the most detailed and comprehensive *social* histories of the everyday uses of technology that we have, and his main argument is that users have shaped the uses of this technology instead of technology shaping society. But Fischer goes too far when he argues that the telephone did not fundamentally change American society. To elaborate my disagreement with him, I would like to begin by conceding his points that the telephone did not greatly expand the relations with distant others, that it reinforced sociability on a local level, and that it led neither to more social isolation nor to greater overall social "connectedness."

The shortcoming of Fischer's argument can be seen if we consider that the telephone has led to an increased frequency of contact with others and to a greater diversity of the circumstances in which—and purposes for which—contact is made. While the telephone may therefore not have transformed American society in terms of the networks of social relationships, there have been more mundane changes; namely, that using the telephone has added to and complemented the existing ways to keep in touch, make arrangements, and express emotions (Nye 1997: 1083), and thus made "aspects of daily life more convenient" (Fischer 1992: 267), and, as Fischer himself points out, "expand[ed] the volume of social activity, and, in that way, add[ed] to the pace of social life" (1992: 254).

Among the aspects that Fischer ignores are the content and diverse communication needs in relationships. And in this case, although it is difficult to pinpoint aggregate changes, it is nevertheless possible to say that an important tool—perhaps *the* most important tool for practicing conviviality—has been added that has allowed individuals to cope with the increasing frequency and diversity of circumstances in which contacts need to be maintained in a more complex society (in a Durkheimian sense) of denser and "thinner" social ties. (I am aware of the circular, "functionalist" logic of explanation of this argument and will return to it shortly.) Thus the number of personal telephone calls per day, both local and long distance, has increased markedly in recent decades. Putnam, for example, cites a recent study that "reported that two-thirds of all adults had called a friend or relative the previous day 'just to talk'" (2000: 166).

All this can be put differently: Fischer may be right to criticize those who argue that the telephone changes people's social relationships. But what his view—"that telephone calling solidified and deepened social relations" rather than changing them (1992: 266, see also 262)—fails to take into consideration is that this technology has become an additional means to cope with the greater frequency and more diverse forms—the greater "complexity"—through which social relationships can and need to be maintained. As Wellman (who, like Fischer, adopts a social network perspective) has argued, our ties have not diminished or become more global, but have rather become more "multiplex," both denser and "thinner," and going beyond instrumental and material needs: "community ties have become ends in themselves, to be enjoyed in their own right and used for emotional adjustment in a society that puts a premium on feeling good about oneself and others" (Wellman 1999: 33). Hence the continuing expansion in frequency, number of contacts, and types of calls in the daily uses of the telephone, which adds to and complements previous communications media (just as newer communications media add to and complement telephony), and the not insignificant amount of time (4.4 hours per week out of 39.6 hours of free time) that is spent on home communication in America (Robinson and Godbey, 1997: 125).

With this, we can briefly summarize some of the cultural changes that relate to technology: one is that the daily social world of the nonelite population before the Industrial Revolution was largely local, whereas it nowadays reaches beyond the local in terms of access to places (cars), sources of mass media content (television), and connections with people (telephone). This fits well with a broader pattern noticed by historical sociologists that has occured over the course of the twentieth century, away from the local to the national level: "National education systems, mass media, and consumer markets are still subverting localism and homogenizing social and cultural life into units which are, at their smallest extent, national" (Mann 1993b: 118).

Second, whereas leisure had previously not been segregated in time and space from the rest of everday life, this segregation is now firmly in place. In this context, a brief comment about the place of everyday consumption is called for, and this can be done by considering the distinction between the public and private spheres: Some have argued that a "privatization" of the household has taken place, and others, conversely, that the private household has been penetrated by the public sphere. Without going into either of these arguments, for our purposes it is sufficient to note that these two processes are compatible with the "segregation" of leisure that I have stressed here—as long as we bear in mind that homes can provide a separate space for leisure, and that a separate time for leisure can be carved out, while the "traffic" between the private household and the realm outside the home can simultaneously increase. In this context we can take note of Nye's argument that the car has enhanced the separation between work, home, and leisure (1998: 243). In a similar fashion, Silverstone has argued that television maintains—rather than blurs—the separation between everyday life and the "non-everydayness" of television viewing (1984: 169).

Again, this segregation of leisure in connection with consumer technologies is part of a larger pattern which Collins, writing from the perspective of historical sociology, describes as follows:

> The modern organization of life into private places, work places, and public places in between them is a historically recent development . . . The realm of consumption is now separated from the places where production takes place and politically- and economically-based power relations are enacted. Consumption now takes place in private or at least outside of situations where it is marked by socially visible rank. The center of gravity of daily life switches to the realm of consumption. This is reinforced by the growth of consumer industries, including entertainment and the hardware which delivers it, into the largest and most visible part of the economy. (2000: 37)[10]

And in America and in Sweden, the visibility of the automobile and information and communication technology industries has of course been particularly "visible."

The Technology-Culture Spiral and the Rubber Cage

How can we make sense of the myriad changes that have accompanied the three technologies over the course of time? In relation to the role of technology in everyday life, Braun distinguishes between the rationalization and the culturalization theses: "whereas rationalization theses emphasize the tendency to standardize everyday life behaviour, and the convergence of the everyday world and the working world resulting from this standardization, culturalization theses stress the diversification in behaviour and the resulting dissociation from the working world" (1994: 100). Braun, and also Joerges (1988), promote the culturalization thesis, or what they call a "technology-culture spiral" of increasing complexity whereby more diverse behavior leads to new technologies; these new technologies result in an integration of several previously different behaviors; this in turn opens up options for new and more diverse forms of behavior, which, in its turn, promotes the use of further technologies; and so on.[11]

This idea, derived in part from the theory of "social differentiation" of Luhmann (whom we encountered in Chapter 5), suggests that the role of technology in everyday life is to lead to increasingly diversified ways of life.[12] Braun and Joerges focus mainly on the immediate social context, and thus on the aquisition of ever greater competencies in the use of domestic technologies. Their "technology-culture" spiral deals with individual social action and not with more long-term changes and more macrochanges. But despite the focus on action, their culturalization thesis and technology spiral idea seem to fit the material presented here: our ways of life have become more complex with the greater use of technologies, and so we use more technologies to cope with this greater complexity. In short, there is a proliferation of technologically mediated cultural activities.

In my view it is useful to apply the Braun/Joerges theoretical perspective in relation to the material presented here, *as long as* we keep in mind the larger frame that puts this increasing diversity or pluralization into context—above all, the scarcity of resources, time, and money, which put an upper limit on the consumption of technologically mediated leisure pursuits, and thus on the uniformity or diversification (again, from a comparative-historical perspective) of a lifestyle of consumption throughout industrialized societies. In other words, we need to go beyond a "social action" perspective and put the culturalization/ technology spiral into the larger context of long-term social change.

Here we can come back to the earlier "functionalist" logic of the argument (that was mentioned earlier in connection with my argument against Fischer's ideas about the telephone): perhaps in the consumption of technology there has been a Luhmann-like (functionalist) evolution toward greater complexity in the sphere of culture—to parallel Beniger's (1986) revolution in "control" and Chandler's (1990) extension of "scale and scope" in the economic sphere and Dandeker's (1990) growth of "surveillance" in the political sphere.[13] Such an "evolution" makes sense of the way I have described the proliferation of technologically mediated cultural activities—on the side of how comparative history or sociology need to "add up" the various diffuse microcontexts into larger macrochanges, and on the substantive side how these changes consist of "more" (paralleling scientific and technological "advance"). This would explain the greater diversity of consumer technologies, their proliferation, and the development of user competencies, and the greater reach, spending of more time, and greater density of ties.

But here we also want to be careful: the greater "complexity" in everyday life, after all, remains confined to the orbit of the household or of private life. So that although this change in everyday life generates more "needs" on the level of the "system" (especially from large technological systems and from the economic system), there remains a disjunction between the micro- and the macro- here. Everyday life remains just that—a way of life—without larger social repercussions except for the forward creep of increasingly complex social needs on the macrolevel. This is why the Luhmann/Braun/Joerges view of the evolution of social complexity needs to be put into a more limited (everyday life) context. This limitation of everyday life is also why the Giddens/Castells view, whereby macrochanges translate directly into the "disembedding" of the social actor, and vice versa, the new disembedded "reflexivity" of social actors which becomes directly relevant to (for example political) macrosocial changes, is (literally) misplaced.[14]

This is perhaps the most important balance that needs to be struck in relation to our topic: that the ownership of cars, television sets, and telephones have become universal in late modern societies, with all the general consequences that have been described—greater mobility, denser and "thinner" contacts, and the domination of time by broadcast (and recently perhaps "narrowcast") entertainment—and that this use of technologies represents neither more freedom or constraint, nor greater disenchantment or reenchantment per se—but a more uniformly diversified consumption lifestyle. This way of life is the product of a society in which the consumption of technology—or technologically mediated cultural activity—has come to occupy the largest share of culture, and this role of technology and way of life are common to all industrialized societies—and they are exclusively modern.[15]

If we come back to the relation between technology and culture then, this argument implies both rationalization (uniformity) and culturalization (increasing diversity), but it implies that the cultural significance of technology is that the microchanges in modern everyday life that have been documented here add up to larger macrochanges. The truth in the "rationalization" thesis that Braun and Joerges overlook is that disenchantment *does* take place if we mean by this the progressive displacement of a nontechnological culture by a culture that is technologically mediated, on the micro- as well as (if we aggregate the micro-) the macrolevels. Perhaps in relation to consumption this is simultaneously disenchantment and "reenchantment"—a "rubber" as much as an "iron cage" (Gellner 1987: 152). The cage is rubber inasmuch as it leaves plenty of scope for comfort or user-friendliness, individuality, choice, and meaningful sociability—but also iron inasmuch as the array of technological mediation between ourselves and our natural and social environments has become a necessity. (This also requires a time perspective in relation to the career of individual technologies: initially, new technologies seem to open up lots of possibilities, but once in everyday use, technologies become routine.)

A different balance to weigh is thus between continuity and change, but this is perhaps best described in terms of a contrast between different disciplinary perspectives: from an anthropological viewpoint, we can say, following Löfgren (1995: 53) and others, that the consumption of cars, television sets (in his case radios, but the same applies to the telephone, as here), and the telephone become routinized after a period of novelty. That is, they become used so routinely that they are taken for granted. From a sociological point of view, on the other hand, we do not want to overlook—or to exaggerate—the social changes that have been brought about by the new technologies. The difference between the anthropological (and historical) perspectives and a sociological one can therefore also be described as follows: the former aim to document the distinctiveness of certain ways of life or patterns of change, whereas the latter aims to provide a systematic account of (ultimately all of) these ways of life and patterns—hence a comparative-historical and synthetic approach in sociology. Sociological insights, in turn, can inform social historians and historians of technology, even if they in no way replace the need for more detailed historical investigations of the role of technology in everyday life.

The argument that I have made here is that the critics of technological determinism are right (on the consumption side) to criticize the widely held belief that technologies have caused profound transformations—*outside* of their social context. But their criticism has also gone too far: the three technologies examined here have led to cumulative changes in social relations and activities; to more mobility in space, to more expenditure (both quantity, and a greater share of) in time and in money on consumption, and to an intensified and more uniformly diversified (to repeat my earlier phrase) pursuit of leisure experiences. In this sense, the argument made here about technology in everyday life fits into the overall argument made in this book, that science and technology *do* shape society, against the social shaping and social constructivist ideas discussed earlier.

If we concentrate not just on the role of technology, but also on the other two terms in the title of this chapter, we can summarize as follows: everyday life has changed in becoming more (uniformly) diversified and more leisure-oriented, with an increasing separation between work and leisure into distinct spheres. Consumption has become more central in social life, with economic growth enabling a greater share of resources to go toward leisure and consumer technologies, and conversely, consumption for its own sake becoming a more central part of the economy. Thus, our circles of the economic and cultural spheres (to come back to Figure 6.1) should overlap more, and this overlapping area (and the triangle of science and technology, and box of mediated culture) should also simply be bigger in size at the end than at the beginning of the twentieth century—the added complication being that, at the microlevel, the consumption of technologies in everyday life also occupies a segregated place. Yet this overlap can be seen as the main macroconnection between spheres, as the earlier quote from Collins (2000: 37) indicated, between a technological culture and the economy, as well as a micro-macro connection, whereby the cumulative changes in everyday life add up to a continual expansion in the consumer economy.

We can now bring the comparative-historical and theoretical parts of the argument more closely together. As we have seen, the patterns of consuming technologies and their role in everyday life have been rather similar—though with time lags—and converged over the course of the twentieth century. This is also a story of "more"—a proliferation of technologies and their roles in everyday life, and an ever wider spread or diffusion among the population.[16] This role, however, is also bounded by everyday life, or by the constraints of existing patterns of time, space, and social interaction. So that although we can speak, with Braun and Joerges, of an increasing mediation of our social life by technology, this mediation needs to be put in the contexts of everyday life and the scope for change within it. Whether the consumption-led economy, including a stream of new technologies, will continue to expand and extend the way new technologies mediate our everyday lives is an open question, but the "freedom" in our more powerful cages and "exoskeletons" will continue to be confined to particular times and places.[17]

7 Who's Afraid of Scientific Objectivity and Technological Determinism?

Summary of the Argument

Before we draw out the implications of the argument, a brief summary will be useful. In this book I have added several concepts to those we already have for understanding the science/technology and society relationship: to science as "representing and intervening," I have added technology as "refining and manipulating" with artifacts that are always also physical. To these definitions I have added a social side by borrowing from Weber the progressive "disenchantment of the world," without going along with his cultural pessimism. To "big science," "large technological systems" and the "system" of mass production, I have added the "rationalization" of production and (following Collins) specified a chain of scientific and technological "advance" from the lab, via mass production, to how a large technological system mediates politics and all the way into the home. I have also, on the consumption side, developed the ideas of Braun and Joerges, and to their "technology-culture spiral," I have added "more," "adding to and complementing" existing technologies, and the increasingly "homogeneous diversification" of how technology mediates everyday life.

To Collins's "high-consensus rapid-discovery science," Fuchs's account of the universality of scientific communication, and Whitley's "task uncertainty" and "mutual dependence" in the production of knowledge, I have added a separation between a cumulative, unified, and unidirectional scientific and technological advance against a multistranded and diffuse culture. I will also shortly add some considerations on the nature of power in relation to scientific and technological advance—and have already added my own "tightening" and "systematic harnessing" to the relationship between technological innovation and economic change. More generally, and aside from individual concepts, I have argued that it is essential to have a comprehensive theory of the science/technology and society relationship, a theory that goes beyond individual studies but also beyond the sociology *of* science and technology as such, as well as beyond theories that submerge science and technology in social or cultural reductionism, and which clearly delimits their autonomy from these other parts of society while showing how they relate to social change.

I have also argued that science and technology must be separated from culture, and at the same time that they cannot be analyzed independently of their effects; or, to put it differently, they must always be related to their causes and consequences in the other spheres of life. This follows from the realist and pragmatist definitions of science and technology, which emphasize the transformation of the physical world and the human-made environment: what science and technology *do*. Further, I have suggested that there is an asymmetry in explaining the role of science and technology that makes them different from explaining other social phenomena—science (and technology) are "irreducible": we can explain some of the sources of the growth of scientific knowledge, but not, ultimately, *why* it "works." We *can*, on the other hand, give an account of *how* science and technology work in producing social change and analyze their consequences—but only with sufficient hindsight, and not (exhaustively) at the research front. In this case, again, there has been a lopsidedness to existing explanations: they have either reduced science and technology to other social forces or been determinist in a speculative way, whereas I have argued that it is the coupling between science and technology and the natural and social worlds that needs explaining.

I have tried to operationalize these concepts and presented a comparative-historical analysis of this "advance" and specified the phases in which the above concepts apply (the major phases are spelled out in summary form in Table 7.1). Finally, the concepts and patterns of change that have been described fit together—scientific and technological advance is a separate part of society whose scope in relation to the other spheres of social life has grown, displacing the (zero-sum) parts of culture and becoming more closely integrated with—or harnessed by—the economic sphere. To Weber's "iron cage," I have therefore added an expanding environmental "footprint" caused by

TABLE 7.1 Key stages in the science, technology, and society relationship

	Science and technology	*Relation to economy and growth*
1600s	Rapid discovery science	
1800s	Great divergence Machines yoked to energy	Exceptional economic growth
1850s	Science technology merger R&D labs Large technological systems	R&D harnessed to economy
1900s	American "system" of mass production	Systematic R&D coupling to economy
1930s	Big science	
1950s	American system diffuses	Systematic harnessing of R&D diffuses
1970s	End of golden age of innovation and growth	

scientific and technological advance, creating an "exoskeleton" as well as a cage (that can be "iron" or "rubber"). This advance remains dynamic at the leading edge, but it has also become sedimented or settled in everyday life. Since the various parts of this overall account also cover the main areas of science and technology and social change, I hope that I have come some distance toward my aim of a comprehensive theory (where "theory" needs to mean nothing more than that concepts fit social change and that the mechanisms of social change are specified—we will come back to this shortly).

Convergence and Advance

In view of the concepts and arguments put forward here, we can return to some of the paradoxes discussed in the introduction. Technological—and scientific—determinism should not lead (as critics of my ideas have put to me) to passivity in the face of change, but the opposite: *only* if we know what the constraints and possibilities are in the midst of how science and technology determine social change, *only* in this case can we see where we should focus our efforts to change the direction of scientific and technological change. Perhaps the most important implication can be derived from my coupling of scientific and technological to economic change: *if* the drive toward economic growth continues, then scientific and technological advance will also generally continue, in which case we need to identify which part of scientific and technological advance to single out for steering—in the knowledge that an enhancement of caging/exoskeleton and of our footprint will also take place. There is an immediate related implication: if we cut down on overall economic growth, then, other things being equal, we will also be able to achieve greater control in the sense of reducing environmental instability.

On the face of it, my argument about scientific and technological advance *seems* to endorse the process of advance, and seems less inclined to ideas about changing it than do constructivist theories of social and cultural shaping. I would argue the opposite: if we conflate science and technology with culture and society from the start, we can get very little sense of where they can be contained, or where their direction can be changed *without* at the same time changing the *whole* of culture or society. In my argument, in contrast, I have on a number of occasions identified limits, boundaries, and forces constraining this advance, and this provides the possibility for a "realistic" (in both senses of the word, epistemological and practical) assessment of where and how, in an age in which we increasingly recognize limits in our manipulation of the physical world, we are able to channel or steer this advance in such a way as to minimize the social costs and maximize the social benefits of science and technology.[1] Put differently, the "inevitabilism" seemingly implied in my argument applies to the advance of science and technology and its various proliferating domains *generally*—this *general* advance is inexorably moving forward; it is a juggernaut (without any implications of speed, or perhaps crawl), while its specific foci and extensions, as well as the resources devoted to different aspects of this endeavor, do not advance in a uniform way.[2]

Before we pursue the further implications of these arguments in more detail, we can briefly identify other ways in which this approach differs from existing approaches. Historians of technology like Fischer (1992) and Nye (1998) are right when they argue that technology did not *by itself* change—in their case American—society. But, if the argument here is correct, then science and technology does not only, as they claim, merely reinforce existing social or cultural patterns. Science and technology change culture such that they create a *distinctive* disenchanted culture. Since science and technology advance, it is necessary to speak of "stages," or of where science and technology have come to in their penetration of society. Once a certain stage is reached, science and technology will affect societies in a similar way (homogeneity and pluralization), but they leave the preexisting patterns in the other spheres of society (cultural, political, and economic) that are unaffected by science and technology to converge or diverge according to their own logic. Put differently, as disenchantment advances, it diffuses in the same way across societies as a separate process, while culture/everyday life and the other spheres (outside science and technology), where they remain unaffected by this process (enchanted, if you like), may develop in different directions.

This is a key reason why the relation between science, technology, and social change is so difficult to grasp or pin down: it leads to progressively greater homogeneity (this follows from the interlocking of science and technology according to regularities in the physical world) *and* to greater pluralization (these regularities enhance our power in using them in ever more, and more varied, ways). In this way, the homogenizing effect of scientific and technological advance can be reconciled with the greater cultural diversity of "more." The ideas about modernization and convergence presented here revolve around advance, but they do not imply a "cultural lag" behind scientific and technological advance because science and technology only advance in relation to the physical or material world; the rest of culture (and the political and economic spheres) remains plural and advances (or not) separately—except insofar as culture (and the other two spheres) are expressed via technology and thus work in the same way.

The "Theory" of Science, Technology, and Social Change—and Limitations of the Argument

As mentioned earlier, theory does not need to refer to anything beyond the concepts and mechanisms that pertain to social change. Put differently, and with a deliberate echo of the view of science presented here, this is the way that concepts or mechanisms and social reality interlock. A different criticism that I have nevertheless encountered is that my separation of science from culture and other distinctions such as that between science and technology could be made in different ways. The reply here is that some distinctions fit the evidence and advance our understanding better than others. Similarly, the distinctions between the spheres of society—or, if you prefer, the types of social power (Mann 1986)—are essential for any comparative-historical and macrosociological analysis. The only distinction that I have made here that does not follow conventional ones is that between science and culture; but, as argued earlier, the only choice here is between treating the scientifically and technologically mediated parts of culture as a qualitatively *distinct* part *within* a concept of culture that embraces them both, or treating them as separate altogether (see Figure 1.2 in Chapter 1).

The sociology of science and technology has recently been dominated by social shaping and constructivism as schools of thought. But these have focused either on microsocial change or—when macroquestions of comparative-historical social change have been addressed—these schools have been based more on philosophy than on the evidence of historical change. Social shaping and constructivism have asserted a general priority of "the social" or an inseparability of science from society or culture on a general level. By contrast, I have argued that it is essential to identify the relations between them concretely and in terms of macro- or comparative-historical changes.

Thus, for example, if the point of Latour, a prominent constructivist, is that science does not exist in isolation, but only insofar as it makes parts of society into a laboratory (as discussed by Collins 1999: 28), then that accords with the ideas presented here. Latour cannot, however, specify the changes in society that result from science since, according to him, science and culture are indistinguishable, and thus these processes of entwining remain local.[3] As we have seen, however, the laboratory changed in scale and was harnessed to production at a particular stage, and the magnitude of this effect was increased when the laboratory became an institution that was widely copied. This diffusion meant that the laboratory's effect could also extend more routinely into society and reach extensively into everyday social life. Or, since Latour favours the idea of "networks," we can say that these networks extend into society not just locally, as Latour argues, but they span the world. In Collins's words, "the world that once existed only in the immediate vicinity of certain European scientists has now expanded around the globe. That Western technoscience works, in Polynesia or Brazil, does not have to be treated as an abstract epistemological question; it is an empirical, sensuously material, practical pattern of how far certain networks have expanded" (1999: 28). Further, laboratory science works this way "around the globe" because of the autonomy of scientific institutions; explaining *culture* in terms of how it is mediated by technology follows an entirely different trajectory (in terms of the networks of how consumer technologies enter everyday life).

One limitation of the theory of science, technology, and social change presented here is that we can specify the changes brought about by science and technology in the past, but the future is "open" insofar as their advance will reach into new domains. This has a number of implications: one is that the idea that science and technology are always already social is true (insofar as it is true) only for the past, where science-technology and society have interlocked and congealed, so that we can also criticize their consequences. But outside of this congealing, science and technology *as such* are beyond social "reproach," except from within, according to criteria of validity and technical efficacy, or where we might want to specify different social priorities for research. Another implication is that examples that have been written about seeking to show the *intrinsically* political or socially shaped nature of science and technology have missed their mark: if science or technology are, for example, distorted by political aims, this is not an intrinsic feature of science or technology.[4] This does not mean that the often expressed view that scientific and technological advance does not take social factors into account enough is mistaken. The reverse: it is precisely by specifying how science, technology, and society interlock with the physical world that we can identify their harmful and beneficial consequences, and this can also be done for future social development inasmuch as social science has predictive powers.

This brings us to another limit of social science. I have argued that our understanding of science, technology, and social change has been impeded by the fact that social science has only covered certain aspects of this relationship and lacked a comprehensive theory. I have also deliberately avoided the label "sociology of science and technology" because a comprehensive social scientific theory needs to address this particular subject matter and at the same time avoid being limited by the boundaries of a subdiscipline. Along the same lines, I have argued that there are no valid objections to the idea of a comprehensive theory: all a-theoretical conceptions of science and technology *implicitly* presuppose a comprehensive theory or background view of the development of science and technology—better to bring it out into the open where it can be assessed.[5] Social shaping and constructivism have put themselves forward as comprehensive theories, but I have argued that they ignore the comparative-historical evidence about the separability and the reciprocal and interlocking influence of science, technology, and social change. Otherwise the field of overarching theories has been left to popular accounts which I labeled speculative determinism and argued that such "wholesale" ideas of social change are equally misleading.

There are, however, good reasons why this topic has been stuck in such theoretical extremes, and we can now, at the end of the argument that has been proposed to overcome them, identify one part of this elusiveness more closely. A tension that has remained in the account put forward here is that scientific and technological advance is both "functional" and "conflictual"—the two main alternatives in social science—depending on the area that we have examined. For example, the scientific community is both generally cohesive (the norms shared in the scientific community)—and at the same time it may be in conflict with parts of society (over resources, for example). And in some areas—national versus global innovation for example—we have found that the two are not necessarily mutually exclusive—even though debates here are often typically framed in terms of zero-sum competition. Another is that this advance is "diffuse"; science and technology can be sharply delimited as phenomena, but in this case the social change brought about by them is diffuse if we think, for example, of disenchantment or of the consumption of technology in everyday life. And finally, as pointed out earlier, social science may have some predictive powers, but this area is also less predictable in the sense that we cannot imagine what lies over the horizon when science and technology migrate again.

One of the main aims of social shaping and constructivism has been to address questions of economic, political, and especially cultural power. The aim has been to change this power in a progressive direction, but without seeing science and technology as separate from society (instead, they are part of politics or culture, so that politics and culture—not science and technology themselves—can be snuck back into the analysis to promote this change). This book has argued that this conflation should not be made and it has presented science and technology as a separate source of power, advancing and diffusing throughout society. It must be added, however, that the question of power is not resolved here, and this is tied not just to the lack of a resolution of the functionalist versus conflict perspectives, but to the double-edged nature of power. Mann has distinguished between diffused versus authoritative power, where diffused power "spreads in a more spontaneous, unconscious, decentered way throughout the population, resulting in similar social practices that embody power relations but are not explicitly commanded," whereas authoritative power is based on commands and obedience (1986: 8). He also distinguishes extensive and intensive power, the former consists in "the ability to organize large numbers of people over far-flung territories in order to engage in minimally stable cooperation," the latter is more local and tightly organized (1986: 7). We have already encountered (in Chapter 3 with the emergence of large technological systems that provide infrastructures) Mann's distinction between collective and distributive power. Now the balance between these two sides would need to be weighed for different areas of scientific knowledge and technological artifacts, but clearly in the case of scientific knowledge and technological artifacts, the balance lies with collective, diffused, and extensive power. This is not surprising since I have focused on knowledge, economic growth, and consumption—rather than politics, militarism, and work. But the question of nature of scientific and technological power needs to be examined further, especially because of the double-edged nature of power: the combination of caging and exoskeleton in my account straddles this divide.

Moreover, related problems have been noticeable in the substantive account of the topic given here: on the one hand, it has been argued that it is necessary to address large comparative-historical changes about the overall shape of this relationship (which the sociology of science and technology typically ignores), but on the other hand, that when this was done, for example, for the consumption of three major technologies, it was possible to give only a sketchy account of how macrochanges could be seen as a product of the aggregation of microchanges. (It can be added immediately, however, that the macro-micro link remains a problem in social science generally; see, for example, Collins 2000). A related problem is that, as we have seen, there is a tightening of relationship between science, technology, and economic change. This does not mean, however, that the aim of science and technology has become exclusively economic; instead, there is a proliferation of technologies and an expansion of the sphere of scientific knowledge in relation to the other spheres. The tightening *does* mean more routine institutional linkages, such that scientific and technological advances can be economically "exploited" and economic resources regularly devoted to scientific and technological advance. But although this is a functional (and sometimes conflictual) relationship, the "values" of the two spheres do not merge, even though they may often overlap—hence "more." And in the case of the mediation of politics, where we might expect more conflict, I argued that technology has contributed to a more extensive and autonomous media system—so again, the balance between conflict and cohesion is "open" since this system is both more managed and allows scope for greater contestation. In short, the account—"theory"—presented here still leaves much to do, partly because the sociology of science and technology has not turned its most advanced tools onto this subject matter, and partly because the tools themselves are incomplete.

At this point, in the spirit of Popperian falsificationism, it may be useful to pose the question, "What if the central argument of this book were wrong?" This book has put forward strong arguments that revolve around taking science and technology seriously and seeing them as a separate influence on society. If the main arguments were false, the following should obtain:

- scientific and technological advance would not introduce social change independently;
- the modern world would not be discontinuous from the premodern world with respect to scientific and technological advance,
- the analytical separation of science and technology from culture or from society would not make sense; science and technology would inherently be part of culture or of politics or economics (or gendered and the like).

We should be able to produce convincing evidence for these ideas, but I have not found them in the literature on these topics. The burden of explanation should therefore fall on explaining science and technology as part and parcel of the other spheres, especially culture, or on identifying some mechanism by which all the elements in the different spheres operate in the same way.

The main support for such arguments comes from epistemology, but a social science explanation of knowledge (as I have argued) requires that it be based on what knowledge *does,* rather than treating epistemology separate from social forces. The closest *sociological* arguments based on what knowledge does, and against the separation of science from culture, have been put forth by Fuchs (2001) and discussed in Chapter 1. In his case, however, machines, for example, become the hard part of culture, and so, I have argued, a different *kind* of culture. And again, if, with Fuchs, we want to see science and technology as a distinctive or qualitatively different kind of culture, then that is consistent with my argument, except that this part of culture (I would not call it that) needs to be not only hard, as for Fuchs, but also cumulative and universal since only this will yield the implications examined here. There is a wider argument about culture in what has just been said: as with all ideas— whether they be labeled culture, ideology, or knowledge—it is not the label or the ideas in themselves that matter for social science, but what they *do.*

There are also a number of substantive limitations of this study. One is that such a wide-ranging topic requires more depth; many areas have not been covered, and many deserve a much fuller treatment. On the other hand, the point here has not been merely to offer a theory or a programmatic statement. It has been to outline an alternative to two prevailing views that I have argued are mistaken. If I have provided enough conceptual scaffolding and enough evidence for an alternative, then I will have succeeded in this first step. Many aspects have nevertheless not been covered: among them military and medical technologies, science and technology in government (apart from political communication), philosophical and other theories, and many areas in the use or the consumption of technology (again, a relatively new area).

In relation to science and technology, it may *seem* as though the topic is so unbounded or the changes resulting from science and technology so manifold as to rule out a comprehensive analysis. In putting it this way, however, we can recognize that this topic should be no different from others in social science—in other words, the "size" of the topic should not make a comprehensive theory any more elusive than for other parts of society. It is worth mentioning here that political or economic sociologists, for example, would not rule out presenting a comprehensive or global theory of social change with equally complex topics.

This can be looked at differently: I have argued that in general, science, technology and the economy have become more closely coupled. Yet some science and technology is remote from the economy and from everyday life. Where to place science and technology is filled with difficulty, but this topic should not be made more complex than it is. Perhaps it is more difficult to make a complete inventory of changes brought about by science and technology in everyday life than it is for, say, the state's role in everyday life. Equally however, there is no a priori reason to rule out such an inventory. In any case, as in other areas of social science, it is important to focus on "first order questions" (Rule 1997: 45–46) with major social implications in order for social science to be able to make progress. The questions that have been treated here in my view meet this criterion.

One major area that has been left out here must at least be mentioned briefly; the realm of work in everyday life.[6] Work has undergone a shift, from a majority occupied with the production of objects to a majority occupied with the manipulation of meanings. It would be possible, I expect, to document similar changes here as in the realm of (nonwork) consumption—"more," homogenization-differentiation, mediation—that are due to the proliferation of technologies of storage, communication, mobility, and others. These were partly encountered earlier—on the macro- or noneveryday level—in relation to technology and economic change. In the case of work and everyday life it would be necessary to provide a detailed account of specialization and to focus (more than in the case of consumption) on the instrumental/disenchanting side as opposed to the expressive side of "cultural" change (Tilly and Tilly 1998: 155–58; Volti 1992: 122–52 for an overview). And as for "disenchantment," again, this should not be one-sidedly associated with pessimism, or, in this case, alienation or de-skilling.

This last caveat needs to be expressed somewhat differently here: the advance of science and technology in the workaday world is a more narrow process of rationalization—in the sense of attempts to make work more effective—as long as we remember that this includes not only mechanization and the like, but also such "scientific" advances as enhancing psychological motivation, more user-friendly technologies, and others. In other words, the disenchantment of work is both iron and rubber. Instead of homogeneous diversification, we can perhaps speak in this case of a deepening or intensifying specialization, noticing that if this process is uniformly adopted, it also results in convergence. And finally, as we saw in relation to the criticisms of Chandler, the convergence on the process of rationalization may not be on best (or most effective) practice in the capitalist enterprise, but it is nevertheless a process of advance and in this sense unidirectional—it yields more powerful practices, yet even this process is uneven and not inescapable. It is socially shaped in being subject to constraints and opportunities, but as with consumption, this process diffuses throughout developed societies and in this sense consists of uniform social change.

The Role of Science and Technology in Modern Society

It is interesting to reflect, in view of the increasingly homogenous diversification or pluralization[7] of science and technology in everyday life—whether such a pattern also applies to culture *apart from* science and technology? One answer to this question might be: there is a more diverse culture today, if we think of volume. It is also true, however, that culture was more diverse in terms of culturally autonomous regions in preindustrial societies. The implication is that—howsoever we want to see the separateness of culture and science/technology in premodern societies—culture and science/technology in modern industrialized societies *share* the feature that they are more diversified wherever they overlap. Here is a potential starting point for the vexed question of what sets culture in developed societies—with their greater scale and scope and lesser particularity—apart from that of developing ones.

The development of science and technology—unidirectional advance of disenchantment—is different from the pattern of the rest of culture. The implication is that we need to keep a concept of culture with two compartments, or have two separate concepts. This separateness is a prominent feature of developed societies, and despite the increasing harnessing of science and technology for cultural, political, and especially economic purposes, there is no reason why the openness and instability of this advance should cease. On the other side, the implication is that the labels for contemporary society that are premised on the idea that advances in science and technology completely dominate the other spheres—the knowledge society, the information society, and so forth—are misplaced. The realm of consumption, as we have seen, and the world of work (shifting toward the manipulation of meanings), have changed. But there is no reason why a shift in this sphere should take priority over changes in the other spheres of life.

A similar point can be made against those who argue that the science and technology are contributing to a "world polity" (Drori et al. 2003) or to the globalization of culture. Such arguments fall into the trap of functionalism, whereby the "needs" of society lead inescapably toward the worldwide or global spread and deepening of scientific and technological institutions. As we have seen, however, science and technology are also in conflict with other parts of society, in addition to causing growing instability in relation to the natural environment. The rise of a global environmental movement (Hironaka 2003) is just one response to this conflict, and the political conflicts around big science and large technological systems (Chapter 3) and the vicissitudes of economic growth (Chapter 4) are others.

In this context it is worth mentioning that changes in the public's attitude to science and technology are mixed. On the one hand, there are measures indicating continued support throughout developed societies.[8] Equally, however, there is an increasing amount of controversy about science and technology (Collins 1993), which reflects the problems of advance. Despite maintaining its high status, the authority and autonomy of science is not uncontested since the lay public has become more aware of the high cost of research and the costs of our environmental footprint. There is no need to invoke the "deficit" model[9] of the public's lack of understanding (and therefore lack of support) of science and technology, as scientists do, or a cultural shift in society at large, as constructivists do—these changes can be explained, among other factors, by reference to resources, interest groups, and social movements (for example, Bauer and Gaskell 2002).

Utopia and Dystopia—or Control of the Human Footprint?

High-tech enthusiasm pervades popular discourse about technology, and this is often tied to technological determinism. The flip side, fear of new technologies or dystopianism, can also be found in popular discourse, for similarly understandable reasons: if technological determinism is speculative, the consequences seem more overwhelming than they are. In the social sciences, on the other hand (and as discussed in Chapter 1), we nowadays often find the opposite, a backlash against the bugbear of technological determinism. Again, part of this is no mystery: many social scientists and others are uncomfortable with the idea that our lives are determined by forces beyond our control. In much recent writing on technology, technology therefore becomes conflated with culture or non-artifact-related social practices, something that takes us away from the determinism or universalism of technology and restores human beings their agency and their cultural particularism.[10]

We have seen why agency or local shaping do not address the constraints and possibilities of scientific and technological advance. A more powerful criticism of scientific and technological determinism has come, as mentioned earlier, not from those who wish to conflate science and technology with culture and society, but from the historian David Edgerton. Edgerton (1998), as mentioned earlier, points out that technological determinism (the same, I would argue, applies to science) is a theory of society, not of technology as such. This is a justified criticism of some theories, but not of the one presented here—since I have always specified the place and limits of science and technology and never considered them in isolation from society and from the work they *do* in society. My arguments have avoided a theory of society driven by science and technology as such because I have specified both the autonomous logic of science and technology and their specific and limited relations to the other spheres (which together add up to society). I have also avoided Edgerton's criticism of an excessive focus on innovation rather than on the actual use of technology and science, and I have not focused exclusively on new technologies but on the importance of some well-established ones.[11]

Edgerton also notes that technology- or innovation-centered views of social change have been part of the self-image of the twentieth century—its myth. I hope I have avoided this myth by means of an analytical and empirical approach, but would add that the Durkheimian reason for the myth is its truth: there have been golden ages of economic growth when it has *seemed* that science and technology were single-handedly responsible for producing this growth and, as we have seen, this is true for certain periods and for major changes (the "great divide," the advent of mass production and consumption, and the diffusion of the American system in the post–World War II period), but not in the overblown way that is commonly thought (the knowledge society and the like, which we will return to shortly). I have argued that it is important to separate the effects of science and technology on economic growth from its effects on everyday life (which *is* culture) and on the other spheres. Each effect, in other words, needs to be "boxed in," put in its proper place, for otherwise the myth will persist.[12]

The argument about science and technology presented here, despite its determinism, is conservative in the sense that the definitions of science and technology stay the same throughout modernity (even if we need to add big science and large technological systems for the twentieth century). Others have argued that the very nature of science and technology have changed.[13] Yet on my view, science and technology have remained the same while politics and culture have changed around them. Examples are the environmental social movement and the possible return to open science after the end of the closed-world part of deterrence-science militarism of the cold war.

As with other concepts, the concepts of science and technology identify one part of social change and do not encompass it wholesale. This brings us to further implications of the theory put forward here: there is no concept of science and technology that could problematize the scientific and technological enterprise as a whole package. As we have seen, at their widest extent, the consequences of science and technology cause "instability," and it is possible to specify the extent of this instability and ways to counteract it. As to individual advances *within* science and technology, there are of course choices about these. Theories which aim at rejecting or endorsing science and technology as a whole, however, belong to areas other than social science.[14] For social science at its broadest, it may be possible to estimate when the impact of science and technology was greatest; perhaps during the second Industrial Revolution or during the postwar Golden Age of growth. Even here, it is necessary to caution that this applies to the contribution of science and technology to growth in the economic sphere and will be different for the other spheres of life.

Disenchantment and greater power over the environment have advanced in step with scientific and technological advance, and this advance entails greater freedoms in some respects (an exoskeleton and a rubber cage) and more constraints (an iron cage of impersonal mediation and greater instability in relation to the environment) in others. Determinism here, as elsewhere in social life, means that we are constrained and enabled. The trajectory of this process should not lead to pessimism or optimism but to a realistic assessment of our options. It is curious that we dislike the feeling of powerlessness in relation to science and technology more than we do elsewhere, but there may be good reasons for this: Scientific and technological advance is coupled to "progress" and our way of life, and its consequences can be seen from a detached or anthropological perspective as a consumer-oriented society with plenty of choice within a framework of continuing economic expansion. Yet, as we have seen, there is also a dark side to science and technology in the sense that our growing power over environment leads to "incalculable" results since our power, or our expanding footprint on the environment, is too complex and uncertain to be evaluated as a whole (hence the scare quotes around "incalculable"). There are different factions in the disputes over these consequences, but one norm that is becoming increasingly well established is that valid scientific knowledge is used to assess the respective claims (Hironaka 2003). Note, however, that one argument that has been used by both sides, "it's not the technology" (or, we might add, the scientific knowledge), "it's how we use them," that was mentioned at the outset of this study, is mistaken. Both constructivists, who think that science and technology is socially shaped, and believers in "pure" science and science that is not shaped by society, subscribe to this view. This is misleading, however, if the argument made here is correct: advance can never be seen in isolation from its consequences, and since it is never entirely shaped by social forces, it is also never entirely within our control.[15]

This book has argued for a new way to think about the relationship between science, technology, and social change. One implication is that research needs to be refocused, away from cultural and social determinism, or arguing that no separation between science and technology and society is possible in the first place, and toward specifying the interlocking between the two sides. And although a broad overview of this interlocking has been given, there are still many gaps left to be filled. One set of implications, however, is normative: if the relationship between science, technology, and social change as I have presented it here is correct, then science and technology constrain and enable us in a manner that is different from the ways in which this relationship has been presented in previous theories, and in a way that is separate and different from other—political, economic, and cultural—institutions. And if it then becomes possible to identify these constraints (and possibilities)—there may, indeed, be an ethical duty toward future generations to do so, a "duty of anticipation" (Partridge 2001: 386–87).

To pursue these normative implications in more detail, it will be helpful, again, to consider the alternatives: among a-sociological scientific and technological determinists, advance inevitably—regardless of social circumstances—brings utopia or dystopia; scientific knowledge and technological artifacts advance in a blanket way, regardless of the social uses to which they are put. These a-social consequences are therefore also described in this case in blanket positive or negative terms. Constructivists and those arguing for social and cultural shaping, on the other hand, construe science and technology as always already part of culture or society, and since they are thus submerged within society, they have no distinct consequences of their own.

On the view presented here, scientific and technological advance disenchants the world, it increases impersonality and mediation. The implication is that what should be analyzed is *where* this advance impinges on the other spheres—adding to or complementing, or conflicting with or reinforcing, the patterns in these spheres. In other words, the analysis should focus on how the advance of this caging is distinct from and influences these other spheres— enabling greater mastery over the world—and how these consequences differ from other political, economic, and cultural patterns of change. This, finally, brings us to norms, since if we identify the sense in which this mastery is not socially neutral (even if science and technology per se *are*), this may lead to greater responsibility for the positive and negative consequences of science and technology *and* social change.

The role of social science is thus to analyze the interlocking of science and technology and social change and criticize past consequences and foreseeable future ones. Since the role of social science is not to criticize science and technology as such, this approach is likely to be countered with the argument (often heard from scientists and engineers) that "science and technology may provide the solutions." The reply to this point, in turn, which follows from my argument here is that only *social* science—not science and technology in themselves—allow an understanding of the parameters of the foreseeable interlocking between science, technology, and social change, and should therefore play a central role in shaping its consequences. The extent to which social scientific knowledge interlocks with the social world, and thus permits such intervention, is of course an open question, which must be answered by reference to the strength of social science as a social institution (see Collins 1994).

The rethinking presented here therefore has important implications for science and technology policy. For example, it is important to say when scientific and technological advance changes society in a hitherto unknown and qualitative way. Unless we identify the separate effects of this advance, it will always be possible to argue that these are not *just* due to science and technology. But if we *can* identify these changes, we can also think about the possible "slippery slopes" that they are leading toward (examples include genetics and biotechnology, nuclear weapons and defensive shields against them, and the like). This theory can thus also be used to improve science and technology, and nature and society—*if* a better environment (which is produced by both) is seen as an aim. Social science can improve society and not just add to cultural hopes and fears as the main alternatives do.

There are more specific normative or policy implications for large technological systems and big science: These systems have a conservative momentum, and more than that they are locked into existing tracks. In this case there is a need to analyze if the direction of the locked-in system continues to fit our needs. This does not mean endorsing, for example, Hughes's (1998) idea that large technological systems in the late twentieth century have changed fundamentally—that they have become more malleable in view of their more complex entwining with society. Nevertheless, in a period in which it is increasingly possible to recognize that big science and large technological systems impose ever greater burdens on society—partly due to their environmental footprint, and partly due to their cost—it becomes more urgent to ask how these juggernauts can be steered toward greater benefits—*given* that they are so difficult to steer.

Apart from these large institutions, there is another way in which science, technology, and social change fits with the broader sociology of developed societies. Science and technology in society are stratified and differentiated in terms of how they affect different social groups. The normative implications of this aspect of social change belong to a wider context of stratification in society (and thus belong to political and economic sociology), even if some of these implications—for example, access to technology, or the accessibility of scientific knowledge—will fall within the realm of the sociology of science and technology and of scientists and engineers. The same goes for the sociology of development, where the contribution of science and technology needs to be weighed against other forces (Chang 2002: esp. 55–58; Hobson 2004; Inkster 1991a: 271–303).

This approach or theory also presents a broader challenge to social science. The argument here is based on the uniqueness of the transformation of the developed world by science and technology. This transformation has costs and benefits. The benefits are too obvious to be listed here, but we can point, for shorthand's sake, to Gellner's argument that the developing world is eagerly embracing industrialization.[16] Yet it is also important to mention the costs; industrialized society, as Crone has pointed out (1989: 196–97), has uniquely embarked on a course of instability.[17] Instability here means permanent social change that sets society on a course where the consequences to the natural and social environment are less controllable, unlike in stable traditional societies. The challenge for social science is therefore to develop better tools to steer this juggernaut. These tools would consist of complex models not just of environmental change (which have emerged in recent decades), but also policies designed to ensure greater stability for the human footprint as a whole, and particularly how the interlocking of science, technology, and society contributes to such stability. Furthermore, the ideas presented here, by avoiding the other two extremes, will lead to a more concrete assessment of the limited but important role that science and technology play in this process.

Apart from providing a theoretical framework that allows social scientific knowledge about science and technology to advance, the payoff from the theory (or approach, if you prefer) to science and technology put forward here is that it allows a realistic engagement with policy questions. The sociology of science and technology has had little such engagement, which has been mainly left to political scientists interested, for example, in funding priorities, or economists trying to promote innovation, or environmental scientists seeking to develop new technologies. Apart from this, it is mainly advocacy groups and social movements that have influenced public opinion and shaped policy. The theory offered here would address this policy engagement in terms of the costs and benefits of the advance of science and technology. At its broadest, the policy implications are to frame and specify the costs and benefits in terms of the enhanced or diminished possibility of control over social life. This avoids, once again, the catchall solutions that there is no independent scientific and technological change that avoids specifying any consequences attributed to science or technology itself, as well as the boosterist view that everything will change without specifying any concrete social consequences. The main advantage of this approach over constructivism is therefore not that it produces a shift in our thinking in the sense of a shift in culture, but rather an understanding of the impact of science and technology *on* culture and politics and the economy, which will in turn allow a constructive engagement with other social scientists and other forces in society—including scientists and engineers.

Notes

Notes to Chapter 1

1. Accessible overviews can be found in Hess (1997), Degele (2002, in German), Weingart (2003, in German), and Volti (1992, less theoretical than Degele). Other surveys of social studies of science and technology include Collins and Restivo (1983), the handbook edited by Jasanoff, Markle, Petersen, and Pinch (1995), and the handbook entry by Knorr Cetina (2005).

2. The preface to the 2nd edition of the most well-known text in this area acknowledges this continuity, MacKenzie and Wajcman 1999 (1st edition 1985): xiv–xv; see also the introductory essay, pp. 21–24. While social shaping was in its heyday in the 1970s and 1980s, social construction came into its own in the later 1980s and 1990s, one prominent text being that of Bijker, Hughes, and Pinch (1987). The argument that the sociology of science should also be applied to technology was made, among others, by Woolgar (1991).

3. One such philosophical idea behind the inseparability of the technological and the social seems to be that nonsocial elements such as nonhuman living things or artifacts can be social actors; see Hess (1997: 83).

4. This argument about the relation between science and economic growth has been made by Gellner (e.g., 1985: 115–16; 1988: 17–18), and he himself has noticed the "pulling oneself by one's own bootstraps" nature of the argument (1985: 89–93). But this will be addressed in Chapter 4. See also Schroeder (1996) for a discussion of Gellner's argument.

5. In other words, one point that mitigates this circularity, and thus also prevents the theory of science, technology, and social change from becoming a theory of society as a whole (something rightly criticized by Edgerton, 1998) is that the unique

change brought about by science and technology in this respect is only a change in the material basis of society—and not in the other, political and cultural realms. Further, this material basis, the growth in output, is, again, a very specific change, not a change in the economic realm as a whole.

6. There is, moreover, an ongoing debate over the sense in which Marx allowed for the autonomous impact of science and technology—if at all (Bimber 1994; MacKenzie 1984).

7. Hacking illustrates the "interlocking of representing and intervening" with various examples from the laboratory sciences. In the case of technological artifacts, examples of the interlocking of refining and manipulating can be found in Hughes's account of how artifacts are made more robust before they leave the laboratory so that they can withstand the tougher conditions in the world outside, and how these conditions, in turn, are imported into the laboratory (1987: 62–63; Collins 1993: 316–17, and Chapters 2 and 3 below).

8. Put differently, technology (and science) have transformed large swathes of the physical and unpopulated environment, which is usually not treated in social science that deals with human relationships, but rather in the environmental sciences.

9. The notion of caging or social caging is also central to Michael Mann's social theory, and it will be developed further later.

10. Shinn and Joerges make an important argument about the intertwining of science and technology by pointing to the role of "research technologies," technologies that drive science since they produce an "objectivization [that] is . . . cumulative and practical" and thus achieves a "practice-based universality" (2002: 244–45). This argument fits well with the ideas put forward here.

11. Christian makes the same point from the point of view of big history: "The most fundamental cultural change of the period [1750–present] was probably the increasing importance of scientific approaches to the world" (2004: 438).

12. Another way to appreciate the extrasocial nature of scientific and technological advance is to ask where the boundaries of the natural as against the social sciences lie. As will be argued later, there are no *intrinsic* boundaries here, but some social scientists, for example, would not want to enter debates about the truth or otherwise about entities where no human (or social) relations are involved—just as natural scientists often do not want to be drawn into certain questions where purely social relations are involved. But these are pragmatic boundaries, not essentialist ones.

Notes to Chapter 2

1. It is of course possible to go to the local or microlevel, to give an account of the local production of individual facts. Latour and Woolgar (1986 2nd ed.) provide the most well-known relativist account of this local process. But only an objective account like Fuchs's, which generalizes and provides an account of how these patterns pertain to a range of settings, and thus embed the micro- in the mesolevel of resources and

institutions, can provide real purchase on this microlevel (1992: 45–76). "Localism" is not relevant here since we are interested in the wider or overall social implications of science and technology, and so in the *advance* of science and technology, which is not local but always translocal and cumulative.

2. Gellner also argues that "culture has become . . . conceptualization minus cognition proper" (1988: 206), separating culture and science or cognition as I do here. Compare Drori, Meyer, Ramirez, and Shofer: "The ethos of modern science is built on the assumption that the world, and the knowledge and scientific laws built on it, have a universal character" (2003: 285).

3. I concede that this is not a direct or one-way process. For example, Cawson, Haddon, and Miles (1995) show the feedback loops whereby research groups in consumer electronics firms, for instance, shape new technologies to fit user needs, partly by seeing how people respond to different new technologies in shops and taking this information back into the laboratory to refine the technologies. Or again, there is the phenomenon of users extending existing science and technology to new applications areas—but, we may ask, is this scientific or technological advance or something else?

4. For the "migration" of scientific disciplines into new areas, see Fuchs (1992: 189).

5. While the social organization of science has been analyzed extensively, the social organization of technology—here, research instruments (but see Shinn and Joerges 2002), engineering disciplines, and the like—have not been systematically examined, so there is little sense of the range of forms or of the organization of technology.

6. These have been described in detail by Collins (1975: 470–523), Whitley (2000), Fuchs (1992), and Becher and Trowler (2001).

7. Edqvist points out that "the system of innovation is larger than the R&D system" (1997: 14, note 37), but R&D systems could also be seen as a wider, supranational effort.

8. There are also supranational efforts to tie these systems together, but the European Union (EU) is the only real example of a systematic effort in this direction, and EU R&D does not yet come close to superseding national efforts (Pavitt and Patel 1999: 114–15).

Notes to Chapter 3

1. See also Christian (2004: 432): "In the nineteenth century, beginning in Germany, science itself began to be incorporated into entrepreneurial activity as companies set up laboratories specifically to raise productivity and profits."

2. Note that the term has a somewhat different meaning for Price (1963), who describes big science in terms of volume—the increase in the number scientists and engineers, in the number of publications, etc. I will follow the usage of the term that has become more common, *big science* as a single large-scale scientific effort.

3. To this it should be added that Westwick sees the national labs as "the foremost manifestation of the new place that science occupies in American society after World War II" (2003: 4).

4. Large technological systems have been mostly national; their main international dimension comes mainly through the coordination of regulation.

5. For these examples, see Mokyr 1990: 115, but before 1850, these are few and isolated technologies, not systems, see 1990: 152.

6. Technology as I define it here; Hughes and followers would not want to make a hard separation between them.

7. One point of Hounshell's essay is nevertheless to argue—unconvincingly in my view (see below)—that Chandler is not a technological determinist.

8. See, for example, Hughes's emphasis on systems as "cultural artifacts," 1983: 461–65. Similarly Latour (1991), who argues that "technology is society made durable": this is misleading since even at this most thorough entwining of technology and society, all analytical power is lost unless the limits of the reach of large technological systems in society are specified. And, as argued earlier, some technologies—from laboratory instruments to consumer durables—are more durable, "harder," parts of culture than others.

9. There have been few attempts to systematically compare how states manage the transformations of the environment on a large scale (Josephson 2004 is an exception).

10. The two countries will be affected by energy politics; Sweden had a strong antinuclear power movement that stopped nuclear power for a time and led to a timetable for phasing out all nuclear energy plants (though this is now being questioned again). This contrasts with the political gridlock on nuclear energy in the United States where environmentalism is one lobbying group among many. This comparison will be taken further below.

11. The emphasis is on comparing large technological systems rather than big science (for Sweden, much would depend on where one wants to draw a line between big and little science). For science, Schott says that "Swedish scientists have performed a small but slightly increasing amount of the world's research and . . . in the global division of labour, Swedish specialisation has emphasized medical science and applied science" (1992: 18). Elzinga describes how research in Swedish universities developed comparatively late, but agrees with Schott that Swedish research nowadays contributes disproportionately—in relation to the size of its economy and population—to global research (1993).

12. Kajser (1999) uses the term *institutional regime* to explain the different national trajectories of large technological systems. Interestingly, he points out that there has been little comparative research between these systems, either between the different systems in one country or between the same systems across different countries.

13. Nye is inconsistent on this point in his account of energy uses in the United States, since he argues that cultural change or change in our way of life can make for the most important changes—he calls these "cultural choices" (1998: 10) or "energy choices" (1998: 263–64). Yet he also notes that changes during the 1970s and 90s were mainly a result of technological ingenuity rather than lifestyle: "Between 1973 and 1993 total energy use increased only 10 percent—an achievement based more on tech-

nological ingenuity than on changes in lifestyle" (1998: 238). But perhaps things have changed since then.

14. There is an important debate about whether military power should be kept separate or subsumed under the political sphere (adding another sphere in Figure 1.1); see the essays by Poggi and Mann in Hall and Schroeder (2006).

15. Compare Hounshell, who says that "what everyone, including those who should have known better, overlooked was that none of these new technologies and products [during World War II] could have emerged without the enormous engineering and manufacturing know-how and capabilities of the nation's corporations . . . Even more overlooked was the degree to which the nation's capabilities in mass production, rather than actions of a group of physicists on a mountaintop in New Mexico or in a laboratory in Chicago, determined the course of war" (1996: 41). The same applies to the cold war: Yes, says Hounshell, "without question, the Cold War drove federal spending for research almost entirely in the direction of the military" (1996: 48). But he also points out that this began to change in the 1960s and 1970s (196: 50), and like Westwick (2003: 300–3), he concludes that the scientific community shaped cold war security concerns to its own purposes as much as it was shaped by them (1996: 57).

16. Weinberger argues that in this instance technology shapes other social forces, yet he does not want to endorse technological determinism. This is curious because the substance of his argument supports technological determinism, which he nevertheless mutes in his concluding sentence where he says that if his argument is "por trayed as technological determinism, it is a socially and politically constructed determinism" (2001: 325), which seems to be having it both ways.

Notes to Chapter 4

1. Gellner (1964: 1–32) discusses the social basis of the idea of permanent progress.

2. For a recent assessment that brings the debate up to date, see David (2000), and Granstrand (1994) for an overview of the debates.

3. One reason for drawing on economic history in what follows rather than on the sociology of science and technology is that, as MacKenzie (1996) has argued, the latter has not engaged with economic change.

4. According to Mokyr, "if revisionist historians such as Pomeranz (2000) are even remotely correct in arguing that the great divergence between Western Europe and the 'Orient' really occurred after 1750, the onus placed on the events we refer to as the Industrial Revolution is all the more weighty" (2002: 285). The same applies to Hobson's (2004) argument: He argues that non-Western science and technology were more advanced and a precondition for "the West's" economic take-off. Yet Hobson, too, concedes a "great divergence" in the nineteenth century. Elman agrees from the point of view of the history of Chinese science: "With the exception of a modernized version of 'traditional Chinese medicine' that flourishes globally as one version of holistic medicine," he says, "the traditional fields of natural studies in imperial China

did not survive the modern science revolution between 1850 and 1920. Instead, Chinese replaced their traditional fields with the modern sciences" (2006: 226).

5. A similar point can be made for science. Jacob, for example, argues that "English science . . . directly fostered industrialization. It was not simply or merely its handmaiden as an older historical tradition once claimed" (1997: 113). Nevertheless, she also argues that English science was a product of culture, a "mental revolution" of "small elites" (1997: 10). Culture was therefore a necessary precondition for science, whereby "British science came wrapped in an ideology that encouraged material prosperity" (1997: 4), and hence the scientific revolution cannot be divorced from its social context (1997: 6). But this argument seems inconsistent (or circular)—to claim a cultural precondition for science *and* simultaneously to argue for its inseparability from the social context. Jacob's argument is more consistent when it suggests a causal link from culture to science, and from science to the Industrial Revolution, which makes sense of her overall claim that "*from a cultural perspective,* by 1815 the Industrial Revolution was over" (1997: 11, italics in the original). But this, in turn, fits with the views of Mokyr and Inkster presented below.

6. S-curves are often used in figures for diffusion whereby after a slow initial start, the gradual shape rising from the bottom of a lying S, there is a sharp rise, which then gradually tapers off until a plateau of adoption is reached (see Rogers 1995).

7. Perhaps in the future we will be able to see that after energy harnessing and machine power, communication—in a wider sense including transportation—was the central shift in the twentieth century whose origins reach back into the nineteenth (Beniger 1986). But we will need more distance to recognize this. We will return to this question in the concluding chapter.

8. We should note that mass consumption only emerges in conjunction with the "American system"; in Mokyr's words: "Only the second Industrial Revolution brought technology to the advantage of the consumer" (2002: 79).

9. See also Mokyr (2002: 152) for the role of communication in the factory system.

10. Van der Wee adds that the American lead over Western Europe and Japan was diminished in this period (1987: 221).

11. Thus in the economic literature, technology is often submerged or internalized in "productivity," in terms of innovations in process versus innovations in product (Rosenberg 1982: 3–8; see also Mokyr 2002: 110), or changes in inputs and outputs—in other words, it becomes a "factor of production" that explains rates of productivity. My emphasis here instead is on social change in production and consumption. This is not to deny that, as Mokyr puts it, when speaking of "the major breakthroughs in the post-1914 decades," that "many of these were . . . improvements upon *existing* techniques rather than totally new techniques" (2002: 108, emphasis in the original).

12. Maddison also offers support (2001: 97) for Mokyr's argument in the first section of this chapter about the fact that the Industrial Revolution ushered in a new era after 1820.

13. Adas argues that "when Americans became increasingly involved overseas, assumptions of their scientific and technological superiority became integral compo-

nents of their version of the civilizing mission" (1989: 406). The same applies to earlier European colonizers. Note however that Adas does not conflate these cultural attitudes with the impact of science and technology per se since his definitions of science and technology and of culture are quite different (see 1989: 5–6 and 9).

14. Perhaps this is why theoretical accounts of scientific and technological advance, especially in economics and economic history, have been so attracted by evolutionism (Mokyr 1990: 327–99). But the advance in science and technology, as we have seen, has not been governed by evolutionary laws, but rather undergone specific historical stages and revolutions.

15. See Chang 2002, for "backwardness" and economic development. Sweden did not just catch up with the American model of innovation. It also had a distinctive trajectory of innovation and economic growth, rooted in its natural endowments before the Industrial Revolution, when timber and mining were at the leading edge of technological and economic development. But in more recent times, although Sweden's developmental path has concentrated on certain sectors, the main difference with America is the state-led development of the mixed economy (Weiss 1995: 83–115; Weiss and Hobson 1995: 95–98).

16. For an overview of innovation in today's Swedish economy, see Sölvell, Zander, and Porter (1993). On a policy level, Glimstedt and Zander (2003) argue that Swedish networks of innovation are not different from other countries like Germany in terms of adopting new technologies. In a similar vein, Jacobsson (2002) argues against the view that the alleged lack of applied research is responsible for Swedish weaknesses in innovation and also against the view that Sweden is unlike America in the levels of funding for research.

17. For a comparison of Sweden's and America's political economy, see Blyth (2002).

Notes to Chapter 5

1. The terms *ICTs* and *media* will be used interchangeably here, depending on whether the context stresses technology or political communication via media, but nothing in the argument hinges on using either term.

2. This criticism has been put most forcefully by Edgerton (1998), who has argued that analyses of technologies tend to focus on innovation and not on actual uses. This is discussed in more detail in Chapter 6.

3. The same point, incidentally, has been made about the analysis of ideology or the worldviews fostered by the media, that these tend to be discussed too much in the abstract without considering their actual effects on political change. In the cases of both countries analyzed here, the evidence about the effects of technology and the effects of media, especially over long periods, is varied. The only defense I can offer for relying on limited evidence is that I have tried to assemble what is available, and I return to some qualifications about how speculative this leaves the argument (in note 21 below). In their introductory essay to a recent volume on comparative political communication research, though it does not deal specifically with *historical*-comparative

research, the editors (Pfetsch and Esser 2004) note that the field is still at a relatively early stage.

4. "Constructivism" is used here in a broader sense than the "constructivism" used in the sociology of science and technology described in Chapter 1. Constructivism as applied in media studies posits that how the media construct our view of the world (and not just scientific knowledge) is a social construction.

5. A recent example is Couldry, who wants to do this by "deconstructing . . . our 'natural' assumption that media are our access points to contemporary social reality," which he calls the "ideology of 'centrality' " (2003: 12), or that there are centers of political and media power. He argues that challenging this assumption will open politics and the media to greater contestation.

6. "Citizen-publics" is used to indicate the broadest sense of people's political engagement with the state, including both informed citizens and the social movements and organizations that are part of "civil society." There has been much debate recently about the role of the state and whether its power has declined in the face of globalization, but Mann (1999) argues that there has been no general decline in the power of the state.

7. Hallin and Mancini (2004) also use the notion of "media systems," but they are mainly interested in political and economic governance, and exclude the defining feature of *technological* systems, which is not only that they consist of interconnected parts, but that they also develop a momentum of their own (Hughes 1994).

8. Ewertsson (2001) also uses the notion of communication as a large technological system, but she only analyzes how it is shaped, and thus on the introduction of the system, rather than the effects of the system on society. Also, she is mainly interested in one part of the system, the new cable and satellite television stations. Silverstone (1984) also applies Hughes's concept to television, and this is discussed in Chapter 6.

9. Other technologies, such as the postal system or the telephone (for campaigning), could be added, but this essay focuses on the main ones.

10. There are, of course, differences in their types of government, chief among them a parliamentary versus a presidential government, but these will not bear on the analysis here.

11. Although Starr (2004) argues that this legal bulwark really only came into force during World War I when the interests of free speech needed to be weighed against the national interest, rather than being an inherent feature of the American system from the start.

12. See, for example, Janowitz's figures from the 1950s to the 1970s, which importantly includes the Vietnam war period (1978: 358). Compare Norris (2000: 300–1) who charts the late 1950s to the 1990s, with some periods in which trust diminished, but no overall decline over the course of this period.

13. It is interesting to note that Rantanen, who compares three families (from Finland, China, and Latvia/Israel) from the end of nineteenth century to the present day, makes the point that, although the media uses of the families gradually drift from the local to the global (though without one displacing the other), "it was the national

that became the most homogeneous" for all three cases (2005: 88) over the course of this period.

14. In international comparisons in the late 1970s, Sweden had more newspapers per population than any country apart from Japan (Briggs and Burke 2002: 214).

15. In the U.S., political consultancy has become a billion dollar industry (Norris 2000: 171–72).

16. Hallin and Mancini also compare a third, "polarized pluralist" model, which applies to Southern and Central European countries.

17. As Hall and Lindholm argue, although interest group politics pervades American politics, the public's view is that politicians should be "above" interest-group politics (1999: 106–7).

18. For "social shaping" versus "technological shaping," see Chapter 1.

19. According to Norris—contra Putnam (2000)—American political activism has remained stable in the postwar period (Norris 2000: 304).

20. See also Cook (2005: 200).

21. In note 3, it was mentioned that the evidence about the effects of technology was sometimes limited. Having reviewed much of what we know about Sweden and the United States over a longer period, it is these *constraints* on what is mediated, or the "scarcity" of the available political communication or of the newshole, and how they have changed, that seems to be the least-well researched area (to my knowledge) and least-well theorized. Put differently, the theoretical focus on the range (and limits) of different media therefore raises questions about what kind of evidence is needed for a complete picture.

22. For the notion of the "affordances" of different media, derived from the work of J. J. Gibson, see Hutchby (2001).

23. Luhmann's theory of communication, based on cybernetics, includes all "differences that make a difference."

24. Luhmann's theory leaves out the role of conflictual role of struggles for visibility in the media, whereby social movements add to the political agenda—yet, as Luhmann might point out, these social movements are also incorporated within the political process, unless they remain a *mere* irritant to it.

25. Bimber notes that "as the marginal cost of information and communication tends towards zero, political associations can form and disband at will" and gatekeeping may be weakened (2003: 104 and 105).

26. This line of thought can be more explicitly linked back to Chapter 3: We can see that this large technological—media—system has been shaped by the political and economic systems in the two countries, even though the technological similarities and their systematic extension are more important than the differences. And again, as we have seen in Chapters 3 and 4, there has been a linear technological development that has extended the scope and scale of this system. This system, moreover, also shapes society; it has become a relatively autonomous vehicle that is used by media and political elites—and, to a lesser extent, by the public—as well as a tool for mediation. This infrastructure has been added to others, a self-sustaining system that delivers ongoing input

to elites and provides output to the public, as well as some feedback in the other direction, and this functioning of the system is common throughout developed societies.

27. The idea of caging, derived from Weber, in relation to technology, has been discussed in Chapter 1. In relation to the modern state, see Mann (1999).

Notes to Chapter 6

1. Cowan's (1987) essay outlined a useful new agenda for research on consumption, but her focus was on the diffusion of new technologies, and not on their uses, as in this chapter. The reason that this chapter is more detailed and pulls together sources about the two countries in a new way is that unlike for the previous chapters, where there has been extensive research and debate, the study of the consumption of technology in everyday life is still ill-defined and there are no studies that try to develop long-term comparisons for several technologies.

2. I would suggest that this chapter can be read as a counterpoint to the "mediation" theme in social studies of technology; that is, one could ask, in the light of my focus on the end uses of technologies, how does the mediation of technologies shape these final uses? On the view presented in this chapter, mediation plays a role primarily in the early stages of adoption, at the outset, and is subsequently "forgotten" once the uses have become routine. Or, as Hughes puts it, "as they grow larger and more complex, [technological] systems tend to be more shaping of society and less shaped by it" (1994: 112). How the American ways of promoting consumption came into place in twentieth-century Europe has been charted by de Grazia (2005).

3. Again, the selection of the two countries is in line with the variation-finding comparative method as described by Tilly (1984: 116–24): Among developed societies, Sweden and America should provide as much of a contrast as one could hope for from a comparative-historical perspective, in terms of their political systems (Chapter 5) and patterns of industrialization (Chapters 3 and 4) and culture (this chapter). They are also quite different, as we will see, for the timing of when the three technologies examined here were introduced.

4. It might be asked why there is no diagram showing science and technology becoming an increasing part—a greater wedge—of the economic and political spheres, as argued in Chapters 4 and 5. There should be, except that saying that science and technology play an increasing role in these two spheres is not controversial, whereas arguing that they play an increasing role in the cultural sphere *is*. Weber and Gellner (1988), as argued in Chapter 1, agree with the position argued here, but I am not aware of others who do so.

5. Braun (1993: 25–26) has also commented on the relative absence of conflict in relation to nonwork everyday technologies, especially once problems of standards and compatibilities at the outset have been overcome. Much of this may have to do with the pluralization or diversification effects of technologies—again, Braun's argument—that will be discussed below. The reason that conflict in the sense that Kline (2000) describes in relation to the *introduction* of the car, for example, strikes us as

"odd" or "amusing" from today's perspective is because of the way in which the car has vanquished this type of conflict so decisively and with such finality.

6. It is important, incidentally, not to equate transparency and "living up to expectations" with conformity. Erickson found, for example, that in relation to material possessions, "alternative" life styles were perhaps more socially acceptable in Munka Ljungby than in Foley (1997: 126–28).

7. Another more general similarity could be mentioned: in both these large and sparsely populated "outlier" countries, nature is something that, more than in the smaller heartlands of earlier industrialization, needs to be transformed by technology (Frykman and Löfgren, 1987: 270).

8. This goes against the view of Flink (1988), who based his ideas that the car contributed to the erosion of sociability partly on *Middletown*.

9. The car and the telephone are good examples of how there are important similarities in the different uses of technology by gender *across* different countries: In Sweden and in America, women drive shorter distances than men (Polk 1997), and they also use the telephone more as a communication technology while men treat it more like an information technology (see Fischer 1992 and Anderson-Skog 1998: esp. 298). This also seems to have applied so far to Internet uses (Haddon 1999 and 2004: 60–68), which would suggest that gender differences in use can also carry *across* different (though related) technologies, though whether Internet uses stay gendered remains to be seen (Brynin 2006 argues they are converging in important respects).

10. One dimension that I have left out here, or mentioned only in passing (but which, I should hasten to add, is also neglected by most of the other studies of this topic, including those that I have drawn on), is the stratification of consumption of technology by class or "status community" (but see Offer 2006: 180–85). All that needs to be said for our purpose, however, is that—apart from the use of new technologies as status display—the *use* of the three technologies in everyday life examined is perhaps stratified mainly along the lines of cosmopolitan vs. local networks, the socio-spatial networks within which everyday social interaction is carried out (Collins 1975: esp. 64–65, 146–52). There continues to be an (unresolved) debate, as in the case of stratification by gender, about whether technology reinforces or weakens this form of stratification.

11. This diversity of new social "actions" makes it hard to pin down the effects of new technologies on the microlevel except in the general terms that I have done here (greater spatial reach, more time spent, frequency of contacts, etc.). The effects of new technologies are diffuse in that television, car, and telephone don't simply allow or constrain the user to do one thing, but many things.

12. Braun and Joerges are close to Luhmann's ideas about "social differentiation," but it needs to be mentioned that Luhmann himself has not theorized the role of technology in everyday life, though he has written extensively about the role of science in society.

13. I mention Dandeker's account of the growth of "surveillance" rather than that of Foucault, who is more well known for this term, because Dandeker provides the basis for a systematic comparative-historical approach to the state, which Foucault does not.

14. As was seen in the previous chapter, the idea that the greater use of information and communication technologies leads to more active citizen-publics needs to be put in the context of the technological system and the limited space that this system affords these publics for engagement.

15. This formulation indicates my disagreement with Latour (1993), who has argued, as the title of his book suggests, that "we have never been modern" in relation to the role of science and technology in society. I am not aware of studies that have made inventories of the main technologies that are used in everyday life across industrialized and industrializing countries, but prima facie there is a lot to be said for the idea that these technologies are very similar in "modern" societies. If we think, for example, of the stock of technologies in a typical household in Sweden and America—not only car, telephone, and television, but also kitchen appliances, computer (perhaps not yet a standard item, but on the way to becoming one) and others, then many of the items are basically the same. But if the items are the same, then so, too, according to the argument here, should their *uses*. Does this imply that the consumption of technology is becoming "globalized"? We cannot be sure, since, as far as I am aware, systematic evidence on the uses of technologies among consumers outside the industrialized world has not been collected in the detail that we have for Sweden, America, and a few other countries. Rantanen's (2005) book discusses globalization and touches on many of the themes in this chapter, especially homogenization. But Rantanen takes a biographical case study approach, which needs to be supplemented by more systematic social history and by arguments that go beyond the use of life histories as illustrations. It is worth stressing here that because technology and culture are kept analytically separate in this essay, the argument does not entail the much-criticized view that cultures will converge or that "culture lags" behind technology: the view presented here does not address culture apart from technology and argues only that those parts of culture that have been transformed by technology become both more homogenous and more diversified.

16. Here the connection with my definition of technology becomes obvious: "refining and manipulating" and "disenchantment" on the side of (cumulative) technological advance translate into "more" on the side of uses, consumption, and the technological mediation of cultural activities. This also means that *more* entails a greater ability to manipulate the social and natural worlds, even if, on the consumption side, technology does not have this instrumental "feel" to it since the main purposes in this case are leisure and sociability.

17. Thus the practical "use" of the study of technology and everyday life that was mentioned in the opening paragraph of this essay (in addition to the non-instrumental value of enhancing our understanding of the cultural significance of technology in modern society) could be to serve as a template for other studies of ongoing technological changes in everyday life comparative perspective, for example Venkatesh's (Shih and Venkatesh 2002) study of the uses of the Internet and other consumer electronics in Sweden and America and Haddon's (2004) work on the mobile phone and Internet in Europe (which includes comparisons with America; see now also Brynin and

Kraut 2006). One question that the study of everyday consumption can shed light on is whether the Internet is likely to be used more like the television or more like the telephone, with all the different consequences for everyday life that this might entail. It can be mentioned in this context that Venkatesh's America-Sweden comparison (Venkatesh 1999; Shih and Venkatesh 2004) and Haddon's five country European study (1999) of the uses of the Internet broadly support one of the arguments made here; namely, that the similarities in everyday uses of the Internet across countries are more striking than the differences. Another set of trends that can be put into context are trends among the younger generation (see Livingstone 2002; Thulin 2004), and whether these are "generational" (pertaining to a younger group) or will prove longer-lasting. It is worth mentioning that for recent decades, we have much better quantitative data for both countries. A few relevant online sources can be mentioned: For Swedish media uses, see www.nordicom.gu.se, which includes annual surveys of daily uses of different media, as well as longitudinal data going back to the late 1970s. The (still ongoing) publication of material from the Middletown study can be followed at www.pbs.org/fmc. The "America's Use of Time Project" can be found at the University of Maryland Survey Research Center at www.bsos.umd.ed/src/. Perhaps the most useful for comparisons of Sweden and America today is the ongoing study by Venkatesh (see Shih and Venkatesh 2004), who has carried out an in-depth analysis of the uses of the Internet and other media in Swedish and American households (and in India), concentrating on consumption, partly published on the project"s homepage (www.crito.uci.edu/noah/publications.htm), and Polk (1997 and 1998), which contains Swedish-American comparisons of mobility. Further, as mentioned earlier, there have been follow-up studies to the Middletown study, which allows very detailed comparisons over time. But there are advantages and disadvantages of (often national) quantitative data, of longitudinal data (again, often national, but see Caplow, Hicks, and Wattenberg 2001 for some data about Middletown), of fine-grained ethnographic studies, and of sociological and historical studies of individual technologies. In my view, the best we can do is to use a mixture of all of these, since grasping the nature of change in everyday life *and* in a comparative-historical perspective *simultaneously* necessitates such a combination of sources.

Notes to Chapter 7

1. This is similar to the argument made by Offer (2006) about reducing economic growth in order to achieve greater well-being. Christian argues along similar lines, asking whether consumption can be "stabilized" (2004: 476). He suggests that "attitudes are critical. The widespread belief that continued growth in production is a good in itself poses one of the main barriers to reform." (2004: 480, see also Josephson 2004: 229–37).

2. Put differently, there seem to be no inherent limits or contradictions or signs of slowing in the growth of scientific knowledge or technological innovation (Christian 2004: 482; Gellner 1988: 265–66), even if we can detect limits to the way they are harnessed to specific goals, the resources required, and the cost to the environment.

3. Where Latour ventures into macrohistorical analysis, he bases his ideas on philosophy rather than on comparative history (see Latour, 1993).

4. See, for a recent example, Joerges's (1999) criticism of Winner's (1980) well-known article about the politics of Robert Moses's bridges.

5. For a similar argument about the philosophy of history or patterns of history generally, see Gellner (1988: 11–12).

6. Even dividing the uses of technology in everyday life into leisure and work, of course, does not exhaust the realm of uses since it leaves out the uses that fall outside both.

7. Diversification means more and different at the same time; pluralization indicates the same thing. The terms can be used interchangeably.

8. Miller and Pardo say that "substantial majorities of adults in Canada, the European Union and the United States continue to hold a positive schema for science and technology, reflecting a positive assessment of the achievements and promise of science" (2000: 125). The attitude is most positive in the United States, with more skepticism in Canada and Europe. There is a more mixed view in Japan which, according to Miller and Pardo, reflects a somewhat lower level of scientific literacy (2000: 125).

9. For this term, see, for example, Durant et al. (2000: 131).

10. It may be that the idea that human behavior and social change are subject to (social) scientific explanation, to disenchantment, is uncomfortable to social scientists and others, and does not allow for enough agency. There is also a counterpart to disenchantment for technology as a cage or exoskeleton: it may be that, as consumers, we have choices about the technologies we use, and comforting to know that we can exercise this control. It is difficult, however, to change whole technological systems and the human-made social environment. We will come back to this below in other contexts.

11. Edgerton (2006) also focuses on established technologies though, since he is an historian, he does not have a theory or overall account of the mechanisms of social change.

12. This includes the golden age of economic growth, which needs to be put in the context that this is when the harnessing of technology to economic growth may have peaked (see Maddison 2001).

13. For science, this view is critically assessed by Whitley (2000: xiii–xxii); for technology, see Hughes (1998).

14. Perhaps they belong to philosophy or normative political science. For science, there is no shortage of literature. For technology, see Mitcham (1994).

15. This is why science and technology often seem to drift or creep forward, as with the argument that is often used that it is impossible to stop scientific and technological advance. It is not impossible, however, to make choices about which individual areas of science and technology should advance, or whether to restrict them—even if the *overall* advance of science and technology continues.

16. "There is a near-universal consensus about . . . growing cognitive wealth, which . . . leads to very powerful technology . . . in deeds rather than in words: those who do not possess such knowledge and technology endeavour to emulate and acquire

it" (Gellner 1985: 91). The same point is made by Mokyr: "Even today resistance to and concerns about technology are still rampant, but the institutional setup of the world is such that holdouts that reject modern technology or cannot adopt it will eventually have to change their minds and somehow limp through the doorway" (2002: 297).

17. "Progress generates a world which is unstable . . . it is kept fragmented because it *wishes* to be unstable, the expansion of cognitive, technological and economic boundaries being its aim" (Crone 1989: 196, emphasis in the original), unlike pre-industrial stable societies.

References

Abercrombie, Nicholas; Hill, Stephen and Turner, Bryan. 1980. *The Dominant Ideology Thesis*. London: George Allen & Unwin.

Adams, Robert McCormick. 1996. *Paths of Fire: An Anthropologist's Inquiry into Western Technology*. Princeton, NJ: Princeton University Press.

Adas, Michael. 1989. *Machines as the Measure of Men: Science, Technologies, and Ideologies of Western Dominance*. Ithaca, NY: Cornell University Press.

Agassi, Joseph. 1985. *Technology, Philosophical and Social Aspects*. Dordrecht: D. Reidel.

Allwood, Martin. 1942. "Lantbrukarna runt Medelby," *Svenska Dagbladet*, March 5.

Allwood, Martin and Ranemark, Inga-Britt. 1943. *Medelby: En sociologisk studie*. Stockholm: Albert Bonniers förlag.

Anderson-Skog, Lena. 1998. "De osynliga användarna: Telefonen och vardagslivet 1880–1995" in Pär Blomkvist and Arne Kajser (eds.), *Den konstruerade världen: Tekniska system i historiskt perspektiv*. Stockholm: Brutus Östlings Bokförlag, 277–98.

Archibugi, Daniele; Howells, Jeremy and Michie, Jonathan. 1999. "Innovation Systems and Policy in a Global Economy," in Daniele Archibugi, Jeremy Howells, and Jonathan Michie (eds.), *Innovation Policy in a Global Economy*. Cambridge, UK: Cambridge University Press, 1–16.

Arthur, Brian. 1989. "Competing Technologies, Increasing Returns, and Lock-in by Historical Events," *The Economic Journal*, 99: 116–31.

Åsard, Erik and Bennett, W. Lance. 1997. *Democracy and the Marketplace of Ideas: Communication and Government in Sweden and the United States*. Cambridge, UK: Cambridge University Press.

Bauer, Martin and Gaskell, George. 2002. "The Biotechnology Movement," in Martin Bauer and George Gaskell (eds.), *Biotechnology: The Making of a Global Controversy*. Cambridge, UK: Cambridge University Press, 379–404.

Becher, Tony and Trowler, Paul. 2001 (2nd ed.). *Academic Tribes and Territories: Intellectual Enquiry and the Culture of Disciplines*. Buckingham, UK: SRHE and Open University Press.

Beniger, James. 1986. *The Control Revolution: Technological and Economic Origins of the Information Society*. Cambridge, MA: Harvard University Press.

Bennett, W. Lance. 2003. "The Internet and Global Activism," in Nick Couldry and James Curran (eds.), *Media Power: Alternative Media in a Networked World*. Lanham, MD: Rowman and Littlefield, 17–37.

Bijker, Wiebe; Hughes, Thomas and Pinch, Trevor (eds.). 1987. *The Social Construction of Technological Systems*. Cambridge, MA: MIT Press.

Bimber, Bruce. 1994. "Three Faces of Technological Determinism," in Merritt Roe Smith and Leo Marx (eds.), *Does Technology Drive History? The Dilemma of Technological Determinism*. Cambridge, MA: MIT Press, 79–100.

Bimber, Bruce. 2003. *Information and American Democracy: Technology in the Evolution of Political Power*. Cambridge, UK: Cambridge University Press.

Blyth, Mark. 2002. *Great Transformations: Economic Ideas and Institutional Change in the Twentieth Century*. Cambridge, UK: Cambridge University Press.

Braun, Ingo. 1993. *Technik-Spiralen: vergleichende Studien zur Technik im Alltag*. Berlin: Edition Sigma.

Braun, Ingo. 1994. "The Technology-Culture Spiral: Three Examples of Technological Developments in Everyday Life," *Research in Philosophy and Technology*, vol. 14 ("Technology and Everyday Life"), 93–118.

Briggs, Asa and Burke, Peter. 2002. *A Social History of the Media: From Gutenberg to the Internet*. Cambridge, UK: Polity Press.

Brubaker, Rogers. 1984. *The Limits of Rationality: An Essay on the Social and Moral Thought of Max Weber*. London: George Allen and Unwin.

Brynin, Malcolm. 2006. "The Neutered Computer," in Robert Kraut, Malcolm Brynin, and Sara Kiesler (eds.). *Computers, Phones and the Internet: Domesticating Information Technology*. New York: Oxford University Press, 84–93.

Brynin, Malcolm and Kraut, Robert. 2006. "Social Studies of Domestic Information and Communication Technologies," in Robert Kraut, Malcolm Brynin, and Sara Kiesler (eds.). *Computers, Phones and the Internet: Domesticating Information Technology*. New York: Oxford University Press, 3–18.

Caccamo, Rita. 2000. *Back to Middletown: Three Generations of Sociological Reflections*. Palo Alto, CA: Stanford University Press.

Campbell, Colin. 1987. *The Romantic Ethic and the Spirit of Consumerism*. Oxford, UK: Basil Blackwell.

Caplow, Theodore; Hicks, Louis and Wattenberg, Ben. 2001. *The First Measured Century: An Illustrated Guide to Trends in America, 1900–2000*. Washington, DC: The AEI Press.

Carey, James. 1989. *Communication as Culture*. London: Routledge.

Castells, Manuel. 2000. "Materials for an Exploratory Theory of the Network Society," *British Journal of Sociology*, 51(1), 5–24.

Cawson, Alan; Haddon, Leslie; and Miles, Ian. 1995. *The Shape of Things to Consume: Delivering Information Technology into the Home.* Aldershot, UK: Ashgate Publishing.

Chandler, Alfred D. 1990. *Scale and Scope: The Dynamics of Industrial Capitalism.* Cambridge, MA: The Belknap Press of Harvard University Press.

Chang, Ha-Joon. 2002. *Kicking Away the Ladder: Development Strategy in Historical Perspective.* London: Anthem Press.

Christian, David. 2004. *Maps of Time: An Introduction to Big History.* Berkeley: University of California Press.

Cole, Stephen. 1992. *Making Science: Between Nature and Society.* Cambridge, MA: Harvard University Press.

Collins, Randall. 1975. *Conflict Sociology: Toward an Explanatory Science.* New York: Academic Press.

Collins, Randall. 1979. *The Credential Society: An Historical Sociology of Education and Stratification.* New York: Academic Press.

Collins, Randall. 1986. *Weberian Sociological Theory.* Cambridge, UK: Cambridge University Press.

Collins, Randall. 1992. "What Does Conflict Theory Predict About America's Future?" *Sociological Perspectives,* 36(4): 289–313.

Collins, Randall. 1993. "Ethical Controversies of Science and Society: A Relation Between Two Spheres of Social Conflict," in Thomas Brante, Steve Fuller, and William Lynch (eds.), *Controversial Science: From Content to Contention.* Albany: State University of New York Press, 301–17.

Collins, Randall. 1994. "Why the Social Sciences Won't Become High-Consensus, Rapid-Discovery Science." *Sociological Forum,* 9(2): 155–77.

Collins, Randall. 1998. *The Sociology of Philosophies: A Global Theory of Intellectual Change.* Cambridge, MA: The Belknap Press of Harvard University Press.

Collins, Randall. 1999. "The European Sociological Tradition and Twenty-First-Century World Sociology," in Janet Abu Lughod (ed.), *Sociology for the Twenty-First Century: Continuities and Cutting Edges.* Chicago, IL: University of Chicago Press, 26–42.

Collins, Randall. 2000. "Situational Stratification: A Micro-Macro Theory of Inequality," *Sociological Theory,* 18(1): 17–43.

Collins, Randall. 2004. *Interaction Ritual Chains.* Princeton, NJ: Princeton University Press.

Collins, Randall and Restivo, Sal. 1983. "Development, Diversity and Conflict in the Sociology of Science," *Sociological Quarterly,* 24: 185–200.

Cook, Timothy. 2005. *Governing with the News: The News Media as a Political Institution,* 2nd ed. Chicago, IL: University of Chicago Press.

Corrigan, Peter. 1997. *The Sociology of Consumption.* London: Sage.

Couldry, Nick. 2003. *Media Rituals: A Critical Approach.* London: Routledge.

Cowan, Ruth Schwartz. 1987. "The Consumption Junction: A Proposal for Research Strategies in the Sociology of Technology," in Bijker, Wiebe; Hughes, Thomas and

Pinch, Trevor (eds.), *The Social Construction of Technological Systems*, Cambridge, MA: MIT Press, 261–80.

Cowan, Ruth Schwartz. 1997. *A Social History of American Technology.* New York: Oxford University Press.

Crone, Patricia. 1989. *Pre-industrial Societies.* Oxford, UK: Blackwell.

Czitrom, Daniel. 1982. *Media and the American Mind.* Chapel Hill: University of North Carolina Press.

Dandeker, Christopher. 1990. *Surveillance, Power and Modernity.* Cambridge, UK: Polity Press.

David, Paul. 2000. "Understanding Digital Technology's Evolution and the Path of Measured Productivity Growth: Present and Future in the Mirror of the Past," in Erik Brynjolfsson and Brian Kahin, (eds.), *Understanding the Digital Economy: Data, Tools, Research.* Cambridge, MA: MIT Press, 49–95.

Degele, Nina. 2002. *Einfuehrung in die Techniksoziologie.* Munich: Fink.

De Grazia, Victoria. 2005. *Irresistible Empire: America's Advance through 20th Century Europe.* Cambridge MA: Harvard University Press.

Djerf-Pierre, Monika and Weibull, Lennart. 2001. *Spegla, granska, tolka. Aktualitetsjournalistik i svensk radio och TV under 1900-talet.* Stockholm: Prisma.

Drori, Gili; Meyer, John; Ramirez, Francisco; and Schofer, Evan. 2003. *Science in the Modern World Polity: Institutionalization and Globalization.* Palo Alto, CA: Stanford University Press.

Durant, John; Bauer, Martin; Gaskell, George; Midden, Cees; Liakopolous, Miltos; and Scholten, Liesbeth. 2000. "Two Cultures of Public Understanding of Science and Technology in Europe," in Meinolf Dierkes and Claudia von Grote (eds.), *Between Understanding and Trust: The Public, Science and Technology.* Amsterdam: Harwood Academic Publishers, 131–54.

Edgerton, David. 1998. "De l'innovation aux usages: Dix theses eclectique sur l'histoire des techniques," *Annales HSS*, 4–5, 815–37.

Edgerton, David. 2006. *The Shock of the Old: A Global History of Twentieth Century Technology.* London: Profile Books.

Edquist, Charles. 1997. "Systems of Innovation Approaches—Their Emergence and Characteristics," in Charles Edquist (ed.) *Systems of Innovation: Technologies, Institutions and Organizations.* London and Washington, DC: Pinter, 1–35.

Edwards, Paul. 1996. *The Closed World: Computers and the Politics of Discourse in Cold War America.* Cambridge, MA: MIT Press.

Elman, Benjamin. 2006. *A Cultural History of Science in Modern China.* Cambridge, MA: Harvard University Press.

Elzinga, Aant. 1993. "Universities, Research and the Transformation of the State in Sweden," in Sheldon Rothblatt and Björn Wittrock (eds.), *The European and American University since 1800.* Cambridge, UK: Cambridge University Press, 191–233.

Erickson, Rita J. 1997. *"Paper or Plastic?" Energy, Environment and Consumerism in Sweden and America.* Westport, CT: Praeger.

Ewertsson, Lena. 2001. *The Triumph of Technology over Politics? Reconstructing Television Systems: The Example of Sweden.* Linköping: Linköping University, Department of Technology and Social Change.

Fischer, Claude. 1992. *America Calling: A Social History of the Telephone to 1940.* Berkeley: University of California Press.

Fligstein, Neil. 2001. *The Architecture of Markets: An Economic Sociology of Twenty-First-Century Capitalist Societies.* Princeton, NJ: Princeton University Press.

Flink, James. 1988. *The Automobile Age.* Cambridge, MA: MIT Press.

Freeman, Chris and Soete, Luc. 1997 (3rd ed.). *The Economics of Industrial Innovation.* Cambrige, MA: MIT Press.

Frykman, Jonas and Löfgen, Orvar. 1987. *Culture Builders: A Historical Anthropology of Middle-Class Life.* New Brunswick, NJ: Rutgers University Press.

Fuchs, Stephan. 1992. *The Professional Quest for Truth: A Social Theory of Science and Knowledge.* Albany: State University of New York Press.

Fuchs, Stephan. 1996. "The New Wars of Truth: Conflicts over Science Studies as Differential Modes of Observation," *Social Science Information,* 35(2): 307–26.

Fuchs, Stephan. 2001. *Against Essentialism: A Theory of Culture and Society.* Cambridge, MA: Harvard University Press.

Fuchs, Stephan. 2002. "What Makes Sciences Scientific?" in Jonathan Turner (ed.), *Handbook of Sociological Theory,* New York: Kluwer Academic/Plenum Publishers, 21–35.

Galison, Peter. 1992. "The Many Faces of Big Science," in Peter Galison and Bruce Hevly (eds.), *Big Science: The Growth of Large-Scale Research.* Palo Alto, CA: Stanford University Press, 1–17.

Galison, Peter. 1997. *Image and Logic: A Material Culture of Microphysics.* Chicago, IL: University of Chicago Press.

Galison, Peter and Hevly, Bruce (eds.). 1992. *Big Science: The Growth of Large-Scale Research.* Palo Alto, CA: Stanford University Press.

Galison, Peter and Stump, David (eds.). 1996. *The Disunity of Science: Boundaries, Contexts, and Power.* Palo Alto, CA: Stanford University Press.

Gans, Herbert. [1979] 2004. *Deciding What's News* (2nd ed.). Evanston, IL: Northwestern University Press.

Gellner, Ernest. 1964. *Thought and Change.* London: Weidenfeld and Nicolson.

Gellner, Ernest. 1979. "Notes Towards a Theory of Ideology," in his *Spectacles and Predicaments: Essays in Social Theory.* Cambridge, UK: Cambridge University Press, 117–32.

Gellner, Ernest. 1985. *Relativism and the Social Sciences.* Cambridge, UK: Cambridge University Press.

Gellner, Ernest. 1987. *Culture, Identity and Politics.* Cambridge, UK: Cambridge University Press.

Gellner, Ernest. 1988. *Plough, Sword and Book: The Structure of Human History.* London: Collins Harvill.

Gellner, Ernest. 1992. *Postmodernism, Reason and Religion.* London: Routledge.

Giddens, Anthony. 1990. *The Consequences of Modernity.* Cambridge, UK: Polity Press.

Glimstedt, Henrik and Zander, Udo. 2003. "Sweden's Wireless Wonders: The Diverse Roots and Selective Adaptations of the Swedish Internet Economy," in Bruce Kogut (ed.), *The Global Internet Economy.* Cambridge, MA: MIT Press, 109–51.

Goldstone, Jack. 2002. "Efflorescences and Economic Growth in World History: Rethinking the Rise of the West and The British Industrial Revolution," *Journal of World History,* 13: 323–89.

Granstrand, Ove. 1994 "Economics of Technology—An Introduction and Overview of a Developing Field" in Ove Granstrand, (ed.), *Economics of Technology.* Amsterdam: Elsevier, 1–36.

Habermas, Juergen. 1982. *Theorie des kommunikativen Handelns.* Frankfurt: Suhrkamp.

Hacking, Ian. 1983. *Representing and Intervening.* Cambridge, UK: Cambridge University Press.

Haddon, Leslie. 1999. "European Perception and Use of the Internet," paper presented at the conference Usages and Services in Telecommunications, Arcachon, France.

Haddon, Leslie. 2004. *Information and Communication Technologies in Everyday Life.* Oxford, UK: Berg.

Hadenius, Stig and Weibull, Lennart. 2003 (8th ed.). *Massmedier—En bok om press, radio och tv.* Stockholm: Albert Bonniers Forlag.

Hagman, Olle. 1998. "Om bilismens utveckling och mening," in Lennart Sturesson (ed.), *Den attraktiva och den problematiska bilismen.* Stockholm: Kommunikationsforskningsberedningen, 31–37.

Hall, John. 2001. "Confessions of a Eurocentric," *International Sociology,* 16(3): 488–97.

Hall, John A. and Lindholm, Charles. 1999. *Is America Breaking Apart?* Princeton, NJ: Princeton University Press.

Hall, John A. and Schroeder, Ralph. (eds.) 2006. *An Anatomy of Power: The Social Theory of Michael Mann.* Cambridge, UK: Cambridge University Press.

Hallin, Daniel and Mancini, Paolo. 2004. *Comparing Media Systems: Three Models of Media and Politics.* Cambridge, UK: Cambridge University Press.

Hess, David. 1997. *Science Studies: An Advanced Introduction.* New York: New York University Press.

Hevly, Bruce. 1992. "Afterword: Reflections on Big Science and Big History," in Peter Galison and Bruce Hevly (eds.), *Big Science: The Growth of Large-Scale Research.* Palo Alto, CA: Stanford University Press, 355–63.

Hironaka, Ann. 2003. "Science and the Environment," in Gili Drori, John Meyer, Francisco Ramirez and Evan Schofer, *Science in the Modern World Polity: Institutionalization and Globalization.* Palo Alto, CA: Stanford University Press, 249–64.

Hobson, John. 2004. *The Eastern Origins of Western Civilization.* Cambridge, UK: Cambridge University Press.

Höijer, Birgitta. 1998. *Det hörde vi allihop! Etermedierna och publiken under 1900-talet.* Värnamo: Fälth & Hässler.

Holt, Douglas. 1988. "Does Cultural Capital Structure American Consumption?" *Journal of Consumer Research*, 25, 1–25.

Horton, Robin. 1970. "African Traditional Thought and Western Science." In Bryan Wilson (ed.) *Rationality*, Oxford, UK: Basil Blackwell, 131–71.

Hounshell, David A. 1984. *From the American System to Mass Production, 1800–1932*. Baltimore, MD: Johns Hopkins University Press.

Hounshell, David. 1992. "Du Pont and the Management of Large-Scale Research Development" in Peter Galison and Bruce Hevly (eds.), *Big Science: The Growth of Large-Scale Research*. Palo Alto, CA: Stanford University Press, 236–61.

Hounshell, David. 1995. "Hughesian History of Technology and Chandlerian Business History: Parallels, Departures, and Critics," *History and Technology*, 12: 205–21.

Hounshell, David. 1996. "The Evolution of Industrial Research in the United States," in Richard Rosenbloom and William Spencer (eds.), *Engines of Innovation: US Industrial Research at the End of an Era*. Boston: Harvard Business School Press, 13–85.

Hounshell, David. 2001. "Rethinking the Cold War; Rethinking Science and Technology in the Cold War; Rethinking the Social Study of Science and Technology," *Social Studies of Science*, 31: 2, 289–97.

Hughes, Thomas. 1983. *Networks of Power: Electrification in Western Society, 1880–1930*. Baltimore, MD: Johns Hopkins University Press.

Hughes, Thomas. 1987. "The Evolution of Large Technological Systems." In Wiebe Bijker, Thomas Hughes, and Trevor Pinch (eds.), *The Social Construction of Technological Systems*. Cambridge, MA: MIT Press, 51–82.

Hughes, Thomas. 1989. *American Genesis: A Century of Invention and Technological Enthusiasm*. Harmondsworth, UK: Penguin.

Hughes, Thomas. 1990. "Walther Rathenau: 'System Builder,' " in Tilmann Buddensieg, Thomas Hughes, Juergen Kocka et al., *Ein Mann vieler Eigenschaften: Walther Rathenau und die Kultur der Moderne*. Berlin: Klaus Wagenbach, 9–31.

Hughes, Thomas. 1994. "Technological Momentum," in Leo Marx and Merritt Roe Smith (eds.), *Does Technology Drive History? The Dilemma of Technological Determinism*. Cambridge, MA: MIT Press, 101–13.

Hughes, Thomas. 1998. *Rescuing Prometheus*. New York: Pantheon Books.

Hult, Jan; Lindqvist, Svante; Odelberg, Wilhelm and Rydberg, Sven. 1989. *Svensk Teknikhistoria*. Hedemora, Sweden: Gidlunds.

Hutchby, Ian. 2001. *Conversation and Technology: From the Telephone to the Internet*. Cambridge, UK: Polity Press.

Inkster, Ian. 1991a. *Science and Technology in History—An Approach to Historical Development*. Basingstoke, UK: Macmillan.

Inkster, Ian. 1991b. "Made in America but Lost to Japan: Science, Technology, and Economic Performance in the Two Capitalist Superpowers," *Social Studies of Science* 21(1): 157–78.

Jacob, Margaret. 1997. *Scientific Culture and the Making of the Industrial West*. New York: Oxford University Press.

Jacobsson, Staffan. 2002. "Universities and Industrial Transformation: An Interpretative and Selective Literature Study with Special Emphasis on Sweden," *Science and Public Policy*, 29 (5): 345–65.

Janowitz, Morris. 1978. *The Last Half-Century: Societal Change and Politics in America*. Chicago, IL: University of Chicago Press.

Jassanoff, Sheila; Markle, Gerald; Petersen, James; and Pinch, Trevor (eds). 1995. *Handbook of Science and Technology Studies*. Thousand Oaks, CA: Sage.

Joerges, Bernward. 1988. "Gerätetechnik und Alltagshandeln. Vorschläge zur Analyse der Technisierung alltäglicher Handlungstrukturen" in Bernward Joerges (ed.), *Technik im Alltag*. Frankfurt am Main: Suhrkamp, 20–50.

Joerges, Bernward. 1999. "Do Politics have Artifacts?," *Social Studies of Science*, 29: 3, 411–31.

Josephson, Paul. 2004. *Resources Under Regimes: Technology, Environment and the State*. Cambridge MA: Harvard University Press.

Kajser, Arne. 1994. *I fädrens spår: Den svenska infrastrukturens historiska utveckling och framtida utmaningar*. Stockholm: Carlssons.

Kajser, Arne. 1999. "The Helping Hand: In Search of a Swedish Institutional Regime for Infrastructural Systems," in Lena Andersson-Skog and Ola Krantz (eds.), *Institutions in the Transport and Communications Industries: State and Private Actors in the Making of Institutional Patterns, 1850–1990*. Canton, MA: Science History Publications, 223–44.

Kargon, Robert; Leslie, Stuart and Schoenberger, Erica. 1992. "Far Beyond Big Science: Science Regions and the Organization of Research and Development," in Peter Galison and Bruce Hevly (eds.), *Big Science: The Growth of Large-Scale Research*. Palo Alto, CA: Stanford University Press, 334–54.

Kline, Ronald. 2000. *Consumers in the Country: Technology and Social Change in Rural America*. Baltimore, MD: Johns Hopkins University Press.

Knorr Cetina, Karin. 2005. "Science, Technology and their Implications," in Craig Calhoun, Chris Rojek, and Bryan Turner (eds.), *The Sage Handbook of Sociology*. London and Thousand Oaks, CA: Sage, 546–60.

Latour, Bruno. 1991. "Technology is Society Made Durable," in John Law (ed.), *A Sociology of Monsters: Essays on Power, Technology, and Domination*. London: Routledge, 103–31.

Latour, Bruno. 1993. *We Have Never Been Modern*. Hemel Hempstead, UK: Harvester Wheatsheaf.

Latour, Bruno and Woolgar, Steve. 1986 (2nd ed). *Laboratory Life: The Construction of Scientific Knowledge*. Princeton, NJ: Princeton University Press.

Lenoir, Timothy. 1998. "Revolution from Above: The Role of the State in Creating the German Research System, 1810–1910," *American Economics Association Papers and Proceedings*, 88(2): 22–27.

Leslie, Stuart W. 1993. *The Cold War and American Science. The Military-Industrial-Academic Complex at MIT and Stanford*. New York: Columbia University Press.

Lindqvist, Svante. 1994. "Changes in the Technological Landscape: The Temporal Dimension in the Growth and Decline of Large Technological Systems," in Ove Granstrand (ed.), *Economics of Technology.* Amsterdam: Elsevier, 271–88.

Livingstone, Sonia. 2002. *Young People and New Media: Childhood and the Changing Media Environment.* London: Sage.

Löfgren, Orvar. 1995. "Consuming Interests" in Jonathan Friedman (ed.), *Consumption and Identity.* Chur, Switzerland: Harwood Academic Publishers, 47–70.

Löfgren, Orvar. 1999. *On Holiday: A History of Vacationing.* Berkeley: University of California Press.

Luhmann, Niklas. 2000. *The Reality of the Mass Media.* Cambridge, UK: Polity Press.

Lundgren, Anders. 1995. *Technological Innovation and Network Evolution.* London: Routledge.

Lynd, Robert S. and Lynd, Helen Merrell. 1929. *Middletown: A Study in American Culture.* New York: Harcourt, Brace.

MacKenzie, Donald. 1984. "Marx and the Machine." *Technology and Culture,* 25: 473–502.

MacKenzie, Donald. 1992. *Inventing Accuracy: A Historical Sociology of Nuclear Missile Guidance.* Cambridge, MA: MIT Press.

MacKenzie, Donald. 1996. "Economic and Sociological Explanations of Technological Change," in his *Knowing Machines: Essay on Technical Change.* Cambridge, MA: MIT Press, 49–65.

MacKenzie, Donald, and Wajcman, Judy. 1999a (2nd ed.). "Preface to the Second Edition," in Donald MacKenzie and Judy Wajcman (eds.), *The Social Shaping of Technology.* Buckingham, UK: Open University Press, xiv–xvii.

MacKenzie, Donald, and Wajcman, Judy. 1999b (2nd ed.). "Introductory Essay and General Issues" in Donald MacKenzie and Judy Wajcman (eds.), *The Social Shaping of Technology.* Buckingham, UK: Open University Press, 3–27.

Maddison, Angus. 2001. *The World Economy: A Millenial Perspective.* Paris: OECD.

Mann, Michael. 1986. *The Sources of Social Power, Volume I: A History from the Beginning to 1760 AD.* Cambridge, UK: Cambridge University Press.

Mann, Michael. 1988. *States, War and Capitalism.* Oxford, UK: Basil Blackwell.

Mann, Michael. 1993a. *The Sources of Social Power, Volume II: The Rise of Classes and Nation-States.* Cambridge, UK: Cambridge University Press.

Mann, Michael. 1993b. "Nation-States in Europe and Other Continents: Diversifying, Developing, Not Dying," *Daedalus,* 122: 115–40.

Mann, Michael. 1999. "Has Globalization Ended the Rise and Rise of the Nation-State?" in T. V. Paul and John A. Hall (eds.), *International Order and the Future of World Politics,* Cambridge, UK: Cambridge University Press, 237–51.

Mann, Michael. 2001. Globalization as Violence. (unpublished essay).

McCracken, Grant. 1988. *Culture and Consumption: New Approaches to the Symbolic Character of Consumer Goods and Activities.* Bloomington: Indiana University Press.

McNeill, J. R. 2000. *Something New Under the Sun: An Environmental History of the Twentieth-Century World.* New York: W. W. Norton.

McNeill, William. 1982. *The Pursuit of Power: Technology, Armed Force and Society Since 1000 AD*. Chicago, IL: University of Chicago Press.

Meyrowitz, Joshua. 1985. *No Sense of Place: The Impact of Electronic Media on Social Behaviour*. Oxford, UK: Oxford University Press.

Miller, Jon D. and Pardo, Rafael. 2000. "Civic Scientific Literacy and Attitude to Science and Technology: A Comparative Analysis of the European Union, The United States, and Japan," in Meinolf Dierkes and Claudia von Grote (eds.), *Between Understanding and Trust: The Public, Science and Technology*. Amsterdam: Harwood Academic Publishers, 81–129.

Mintz, Sydney. 1997. "Swallowing Modernity," in James L. Watson (ed.), *Golden Arches East: McDonald's in East Asia*. Palo Alto, CA: Stanford University Press, 183–200.

Mitcham, Carl. 1994. *Thinking Through Technology: The Path Between Engineering and Philosophy*. Chicago, IL: University of Chicago Press.

Mokyr, Joel. 1990. *The Lever of Riches—Technological Creativity and Economic Progress*. Oxford, UK: Oxford University Press.

Mokyr, Joel. 2002. *The Gifts of Athena: Historical Origins of the Knowledge Economy*. Princeton, NJ: Princeton University Press.

Norris, Pippa. 2000. *A Virtuous Circle: Political Communication in Post-Industrial Societies*. Cambridge, UK: Cambridge University Press.

Nye, David. 1997. "Shaping Communication Networks: Telegraph, Telephone, Computer," *Social Research*, 64(3): 1066–91.

Nye, David. 1998. *Consuming Power: A Social History of American Energies*. Cambridge, MA: MIT Press.

Offer, Avner. 2006. *The Challenge of Affluence: Self-Control and Well-Being in the United States and Britain Since 1950*. Oxford, UK: Oxford University Press.

Orfali, Kristina. 1991. "The Rise and Fall of the Swedish Model" in Antoine Prost and Gerard Vincent (eds.), *A History of Private Life: Riddles of Identity in Modern Times*. Cambridge, MA: Harvard University Press, 417–49.

Page, Benjamin. 1996. *Who Deliberates? Mass Media in a Modern Democracy*. Chicago, IL: University of Chicago Press.

Partridge, Ernest. 2001. "Future Generations," in Dale Jamieson (ed.), *A Companion to Environmental Philosophy*. Oxford, UK: Blackwell, 377–89.

Pavitt, Keith and Patel, Parimal. 1999. "Global Corporations and National Systems of Innovation: Who Dominates Whom?," in Daniele Archibugi, Jeremy Howells and Jonathan Michie (eds.), *Innovation Policy in a Global Economy*. Cambridge, UK: Cambridge University Press, 94–119.

Perkin, Harold. 1996. *The Third Revolution: Professional Elites in the Modern World*. London: Routledge.

Perrow, Charles. 1984. *Normal Accidents: Living with High-Risk Technologies*. New York: Basic Books.

Perrow, Charles. 2002. *Organizing America: Wealth, Power, and the Origins of Corporate Capitalism*. Princeton, NJ: Princeton University Press.

Pfetsch, Barbara, and Esser, Frank. 2004. "Comparing Political Communication: Reorientations in a Changing World," in Frank Esser and Barbara Pfetsch (eds.), *Comparing Political Communication: Theories, Cases, Challenges.* Cambridge, UK: Cambridge University Press, 3–24.

Poggi, Gianfranco. 2006. "Political Power Un-Manned: A Defense of the Holy Trinity from Mann's Military Attack," in John Hall and Ralph Schroeder (eds.), *An Anatomy of Power: The Social Theory of Michael Mann.* Cambridge, UK: Cambridge University Press, 135–49.

Polk, Merritt. 1997. "Swedish Men and Women's Mobility Patterns: Issues of Social Equity and Ecological Sustainability," in *Women's Travel Issues: Proceedings of the Second National Conference,* October 1996, Washington DC, U.S. Department of Transportation, Federal Highway Administration, Office of Highway Information Management, Publication no. FHWA-P2-97-024, 187–211.

Polk, Merritt. 1998. *Gendered Mobility: A Study of Women's and Men's Relations to Automobility.* PhD. Dissertation, Department of Interdisciplinary Studies of the Human Condition, Gothenburg University.

Pomeranz, Kenneth. 2000. *The Great Divergence: China, Europe, and the Making of the Modern World Economy.* Princeton, NJ: Princeton University Press.

Price, Derek J. de Solla. 1963. *Little Science, Big Science.* New York: Columbia University Press.

Price, Derek J. de Solla. 1986. *Little Science, Big Science, and Beyond.* New York: Columbia University Press.

Putnam, Robert. 2000. *Bowling Alone: the Collapse and Revival of American Community.* New York: Simon and Schuster.

Rantanen, Terhi. 2005. *The Media and Globalization.* London: Sage.

Robinson, John and Converse, Philip. 1972. "Social Change Reflected in the Use of Time," in Angus Campbell and Philip Converse (eds.), *The Human Meaning of Social Change.,* New York: Russell Sage Foundation, 17–86.

Robinson, John and Godbey, Geoffrey. 1997. *Time for Life: The Surprising Ways Americans Use Their Time.* University Park: Pennsylvania State University Press.

Rogers, Everett. 1995. *Diffusion of Innovations* (4th ed.). New York: Free Press.

Rosenberg, Nathan. 1982. *Inside the Black Box: Technology and Economics.* Cambridge, UK: Cambridge University Press.

Rosenberg, Nathan. 2004. "America's University/Industry Interfaces 1945–2000", paper presented SCANCOR, Stanford University, Sept.1.

Roy, William. 1990. "Functional and Historical Logics in Explaining the Rise of the American Industrial Corporation," *Comparative Social Research,* 12: 12–44.

Rucht, Dieter. 1995. "The Impact of Anti-Nuclear Power Movements in International Comparison," in Martin Bauer (ed.), *Resistance to New Technology: Nuclear Power, Information Technology and Biotechnology.* Cambridge, UK: Cambridge University Press, 277–91.

Rule, James. 1997. *Theory and Progress in Social Science.* Cambridge, UK: Cambridge University Press.

Schott, Thomas. 1992. "Scientific Research in Sweden: Orientation Toward the American Centre and Embeddedness in Nordic and European Environments," *Science Studies* 5(2), 13–27.

Schroeder, Ralph. 1995. "Disenchantment and Its Discontents: Weberian Perspectives on Science and Technology," *Sociological Review*, 43: 2, 227–50.

Schroeder, Ralph. 1996. "From the Big Divide to the Rubber Cage: Gellner's Conception of Science and Technology," in John Hall and Ian Jarvie (eds.), *The Social Philosophy of Ernest Gellner*. Amsterdam: Editions Rodopi, 427–43.

Schroeder, Ralph. 1997. "The Sociology of Science and Technology After Relativism," in David Owen (ed.), *Sociology After Postmodernism.*, London: Sage, 124–37.

Schumpeter, Joseph. 1934. *The Theory of Economic Development*. Cambridge, MA: Harvard University Press.

Schumpeter, Joseph. [1942] 1994. *Capitalism, Socialism and Democracy*. London: Routledge.

Shih, Chuan Fong and Venkatesh, Alladi. 2003. "A Comparative Study of Home Computer Use in Three Countries: U.S., Sweden and India," CRITO Working Papers, www.crito.uci.edu/noah/publications.htm.

Shinn, Terry and Joerges, Bernward. 2002. "The Transverse Science and Technology Culture: Dynamics and Roles of Research Technology," *Social Science Information*, 41(2): 207–51.

Silverstone, Roger. 1984. *Television and Everyday Life*. London: Routledge.

Silverstone, Roger and Hirsch, Eric (eds.). 1992. *Consuming Technologies: Media and Information in Domestic Spaces*. London: Routledge.

Smith, Merritt Roe and Marx, Leo (eds.). 1994. *Does Technology Drive History? The Dilemma of Technological Determinism*. Cambridge, MA: MIT Press.

Snow, C. P. [1959] 1964. (2nd ed.). *The Two Cultures: And a Second Look*. Cambridge, UK: Cambridge University Press.

Sölvell, Örjan; Zander, Ivo and Porter, Michael. 1993 (2nd ed). *Advantage Sweden*. Stockholm: Norstedts Juridik.

Starr, Paul. 2004. *The Creation of the Media: Political Origins of Modern Communications*. New York: Basic Books.

Stehr, Nico. 1994. *Knowledge Societies*. London: Sage Publications.

Summerton, Jane. 1994. "Introductory Essay: The Systems Approach to Technological Change" in Jane Summerton (ed.), *Changing Large Technical Systems*. Boulder, CO: Westview Press, 1–21.

Thompson, John B. 1995. *The Media and Modernity: A Social Theory of the Media*. Cambridge, UK: Polity Press.

Thörnqvist, Rolf. 2000. "Etnografisk litteratur eller vetenskaplig rapport? Sociologer kartlägger lokalsamhällets förhållanden," in Bengt Erik Eriksson and Roger Qvarsell (eds.), *Samhällets Linneaner: Kartläggning och förståelse i samhällsvetenskapernas historia*. Stockholm: Carlsson, 227–58.

Thulin, Eva. 2004. *Ungdomars virtuella rörlighet. Användning av dator, internet och mobiltelefon i ett geografiskt perspektiv*. Ph.D. thesis, Department of Human and Economic Geography, Gothenburg University.

Tilly, Charles. 1984. *Big Structures, Large Processes, and Huge Comparisons.* New York: Russell Sage Foundation.

Tilly, Chris and Tilly, Charles. 1998. *Work Under Capitalism.* Boulder, CO: Westview Press.

Trigg, Roger. 1993. *Rationality and Science: Can Science Explain Everything?* Oxford, UK: Blackwell.

Turner, Jonathan. 1997. *The Institutional Order: Economy, Kinship, Religion, Polity, Law and Education in Evolutionary and Comparative Perspective.* New York: Longman.

Van der Wee, Herman. 1987. *Prosperity and Upheaval: The World Economy 1945–80.* Harmondsworth, UK: Penguin Books.

Venkatesh, Alladi. 1999. "Some Recent Findings on Home Uses and Experiences of Media and Information Technology," paper presented at Gothenburg University, Dec. 1.

Vilhelmson, Bertil. 1999. "Daily Mobility and the Use of Time for Different Activities: The Case of Sweden," *GeoJournal,* 48: 177–85.

Volti, Rudi. 1992. (2nd ed.). *Society and Technological Change.* New York: St. Martin's Press.

Ward, Ken. 1989. *Mass Communication in the Modern World.* Basingstoke, UK: Macmillan.

Weber, Max. 1948. *From Max Weber: Essays in Sociology.* London: Routledge and Kegan Paul.

Weber, Max. 1978. *Economy and Society: An Outline of Interpretive Sociology.* Guenther Roth and Claus Wittich (eds.). Berkeley: University of California Press.

Weber, Max. 1980. (4th ed. [1921]). *Gesammelte Politische Schriften.* Tübingen: J. C. B. Mohr.

Weber, Max. 1994. *Political Writings.* (eds. Peter Lassman and Ronal Spiers). Cambridge, UK: Cambridge University Press.

Weinberger, Hans. 2001. "The Neutrality Flagpole: Swedish Neutrality Policy and Technological Alliances," in Michael Thad Allen and Gabrielle Hecht (eds.), *Technologies of Power: Essays in Honor of Thomas Parke Hughes and Agatha Chipley Hughes.* Cambridge, MA: MIT Press, 295-332.

Weingart, Peter. 2003. *Wissenschaftssoziologie.* Bielefeld, Germany: Transcript Verlag.

Weiss, Linda. 1998. *The Myth of the Powerless State: Governing the Economy in a Global Era.* Cambridge, UK: Polity Press.

Weiss, Linda. 2003. "Introduction: Bringing Domestic Institutions Back In" in Linda Weiss (ed.), *States in the Global Economy: Bringing Domestic Institutions Back In.* Cambridge, UK: Cambridge University Press, 1-33.

Weiss, Linda and Hobson, John. 1995. *States and Economic Development: A Comparative Historical Analysis.* Cambridge, UK: Polity Press.

Wellman, Barry. 1999. "The Network Community: An Introduction," in Barry Wellman (ed.), *Networks in the Global Village: Life in Contemporary Communities.* Boulder, CO: Westview Press, 1-47.

Westwick, Peter. 2003. *The National Labs: Science in an American System, 1947–1974.*
    Cambridge, MA: Harvard University Press.

White, Harrison C. 1981. "Where Do Markets Come From?," *American Journal of*
    Sociology, 87: 514–47.

Whitley, Richard. 2000. (2nd ed.). *The Intellectual and Social Organization of the*
    Sciences. Oxford, UK: Oxford University Press.

Winner, Langdon. 1980. "Do Artifacts Have Politics?," *Daedalus,* 109: 12–36.

Woolgar, Steve. 1988. *Science: The Very Idea.* Chichester, UK: Ellis Horwood.

Woolgar, Steve. 1991. "The Turn to Technology in the Social Studies of Science,"
    Science, Technology and Human Values 16(1): 20–50.

Yates, JoAnne. 1989. *Control Through Communication: The Rise of System in American*
    Management. Baltimore, MD: Johns Hopkins University Press.

Index

Adams, Robert McCormick, 63
Adas, Michael, 146n13
advance, scientific and technological, 30,
   31–33: as cumulative, 2, 64, 102, 130,
   142n10, 143n1, 152n16; impact on
   economic growth, 3, 5, 6–8, 12, 19,
   27–28, 37, 39, 41, 45–46, 47, 59, 60–
   63, 64–73, 119–20, 122, 123–24, 129,
   131, 133, 135–36, 139, 140, 141nn4,5,
   150n4, 153n1, 154n12; impact on
   everyday life, 3, 5, 8, 13, 20, 24,
   26–27, 51, 63–64, 69, 75, 99–120,
   121, 122, 123, 124–25, 126, 128, 129,
   131–32, 133, 135, 140, 143n3, 147n2,
   150nn1,2,4,5, 152nn15,16,17; impact
   on leisure and social activities, 99,
   106, 107, 111–12, 115–16, 118, 119–20,
   151n8, 152n16, 154n6; impact on
   political sphere, 2–3, 15–16, 20, 38,
   46, 73, 74–98, 100, 102, 117, 121,
   124–25, 129, 133, 140, 149n25, 150n4,
   152n14; impact on society, 1, 2, 6–8,
   9–12, 15–16, 18–19, 20, 23–25, 30–31,
   50, 56–57, 99–120, 121–33, 134–39,
   145n16, 149n26, 151nn8,10,11,
   152n14,15,16,17, 154nn10,15,16,
   155n17; impact on work, 131–32,
133; instability caused by, 123–24,
   133, 135–36, 139, 154n17, 155n17;
   institutional bases of, 19, 25–33,
   36, 37–38, 62; limits to, 49–50, 56,
   92–93, 96, 97, 117, 120, 124, 133–34,
   136, 149n21, 153n2, 154n15; national
   systems of innovation, 38–39, 42,
   43–44, 70–71, 128, 143nn7,8,3;
   policies regarding, 5, 28, 37, 38–39,
   54, 60, 70, 79, 138, 139–40, 147n16,
   154n15; public attitudes toward,
   133–34, 154n8, 154n16; relationship
   to disenchantment, 2, 9–10, 12, 13,
   17, 31, 35, 41, 62, 63, 67, 103, 118,
   121, 124–25, 128, 132, 133, 136, 137,
   152n16, 154n10; relationship to
   environmental transformation, 13,
   17–18, 46–47, 48–49, 52–53, 54, 67,
   122–24, 133, 136, 138, 139, 142n8,
   144n9, 153n2, 154n10; relationship
   to innovation, 36–38; relationship to
   militarism, 28, 39, 42, 45, 46, 57–59,
   71, 72, 131, 135, 145n15; and social
   power, 122, 125, 128–29
Agassi, Joseph, 8
Allwood, Martin: *Medelby,* 84, 105–8,
   109–10